T3-BOJ-441

Mathematical Problems of Control Theory

An Introduction

SERIES ON STABILITY, VIBRATION AND CONTROL OF SYSTEMS

Series Editors: A. Guran, A. Belyayev, H. Bremer, C. Christov & G. Stavroulakis

---

About the Series

Rapid developments in system dynamics and control, areas related to many other topics in applied mathematics, call for comprehensive presentations of current topics. This series contains textbooks, monographs, treatises, conference proceedings and a collection of thematically organized research or pedagogical articles addressing dynamical systems and control.

The material is ideal for a general scientific and engineering readership, and is also mathematically precise enough to be a useful reference for research specialists in mechanics and control, nonlinear dynamics, and in applied mathematics and physics.

Selected Volumes in Series B

Proceedings of the First International Congress on Dynamics and Control of Systems, Chateau Laurier, Ottawa, Canada, 5–7 August 1999
Editors: A. Guran, S. Biswas, L. Cacetta, C. Robach, K. Teo, and T. Vincent

Selected Topics in Structronics and Mechatronic Systems
Editors: A. Belyayev and A. Guran

Selected Volumes in Series A

Vol. 1 Stability Theory of Elastic Rods
Author: T. Atanackovic

Vol. 2 Stability of Gyroscopic Systems
Authors: A. Guran, A. Bajaj, Y. Ishida, G. D'Eleuterio, N. Perkins, and C. Pierre

Vol. 3 Vibration Analysis of Plates by the Superposition Method
Author: Daniel J. Gorman

Vol. 4 Asymptotic Methods in Buckling Theory of Elastic Shells
Authors: P. E. Tovstik and A. L. Smirinov

Vol. 5 Generalized Point Models in Structural Mechanics
Author: I. V. Andronov

Vol. 6 Mathematical Problems of Control Theory: An Introduction
Author: G. A. Leonov

Vol. 7 Vibrational Mechanics: Theory and Applications to the Problems of Nonlinear Dynamics
Author: Ilya I. Blekhmam

SERIES ON STABILITY, VIBRATION AND CONTROL OF SYSTEMS

Series A Volume 4

Series Editors: **Ardéshir Guran & Daniel J Inman**

Mathematical Problems of Control Theory

An Introduction

Gennady A. Leonov

Department of Mathematics and Mechanics
St. Petersburg State University

Published by

World Scientific Publishing Co. Pte. Ltd.
P O Box 128, Farrer Road, Singapore 912805
USA office: Suite 1B, 1060 Main Street, River Edge, NJ 07661
UK office: 57 Shelton Street, Covent Garden, London WC2H 9HE

British Library Cataloguing-in-Publication Data
A catalogue record for this book is available from the British Library.

MATHEMATICAL PROBLEMS OF CONTROL THEORY: An Introduction

ISBN 981-02-4694-3

Printed in Singapore by Mainland Press

Preface

It is human to feel the attraction of concreteness. The necessity of concreteness is most pronounced in the mathematical activity. The famous mathematician of the 19th century Karl Weierstrass has pointed to the fact that progress in science is impossible without studying concrete problems. The abandonment of concrete problems in favor of more and more abstract research led to a crisis, which is often discussed now by the mathematicians. The mathematical control theory is no exception. The Wiener idea on a partitioning of a control system into the parts, consisting of sensors, actuators, and control algorithms being elaborated first by mathematicians and being realized then, using the comprehensive facilities of electronics, by engineers, was of crucial importance in the making of cybernetics. However at the same time it has been considered as a contributary factor for arising a tendency for an abandonment of concreteness.

The notorious and very nontrivial thesis of I.A. Vyshnegradsky: "there is no governor without friction", and the conclusion of V. Volterra that an average number of preys in the ecological predator-pray system turns out to be constant with respect to the initial data are a result of the solutions of concrete problems only.

The study of concrete control systems allows us to evaluate both the remarkable simplicity of the construction of a damper winding for a synchronous machine, which acts as a stabilizing feedback, and the idea of television broadcasting of two-dimensional pictures via one-dimensional information channels by constructing a system of generators synchronization.

In this book we try to show how the study of concrete control systems has become a motivation for development of the mathematical tools needed for solving such problems. In many cases by this apparatus far-reaching generalizations were made and its further development exerts a material effect on many fields of mathematics.

A plan to write such a book has arisen from author's perusal of many remarkable books and papers of the general control theory and of special control systems in various domains of technology: energetics, shipbuilding, communications, aerotechnics, and computer technology. A preparation of the courses of lectures: "The control theory. Analysis" and "Introduction to the applied theory of dynamical systems" for students of the Faculty of Mathematics and Mechanics, making a speciality of "Applied mathematics" was a further stimulus.

An additional motivation has also become the following remark of K.J. Åström [5]: "The introductory courses in control are often very similar to courses given twenty or thirty years ago even if the field itself has developed substantially. The only difference may be a sprinkling of Matlab exercises. We need to take a careful look at our knowledge base and explore how it can be weeded and streamlined. We should probably also pay more attention to the academic positioning of our field".

The readers of this book are assumed to be familiar with algebra, calculus, and differential equations according with a program of two first years of mathematical departments. We have tried to make the presentation as near to being elementary as possible.

The author hopes that the book will be useful for specialists in the control theory, differential equations, dynamical systems, theoretical and applied mechanics. But first of all it is intended for the students and postgraduates, who begin to specialize in the above-mentioned scientific fields.

The author thanks N.V. Kuznetsov, S.N. Pakshin, I.I. Ryzhakova, M.M. Shumafov, and Yu.K. Zotov for the patient cooperation during the final stages of preparation of this monograph.

The author also thanks prof. A.L. Fradkov and A.S. Matveev who read the manuscript in detail and made many useful remarks.

Finally, the author wishes to acknowledge his great indebtedness to Elmira A. Gurmuzova for her translating the manuscript from Russian into English.

leonov@math.spbu.ru

St.Petersburg, April 2001.

Contents

Chapter 1

The Watt governor and the mathematical theory of stability of motion

1.1 The Watt flyball governor and its modifications

The feedback principle has been used for centuries. An outstanding early example is the flyball governor, invented in 1788 by the Scottish engineer James Watt to control the speed of the steam engine. The Watt governor is used for keeping a constant angular velocity of the shaft of the engine (in a classic steam engine, in a steam turbine or hydroturbine, in a diesel device, and so on).

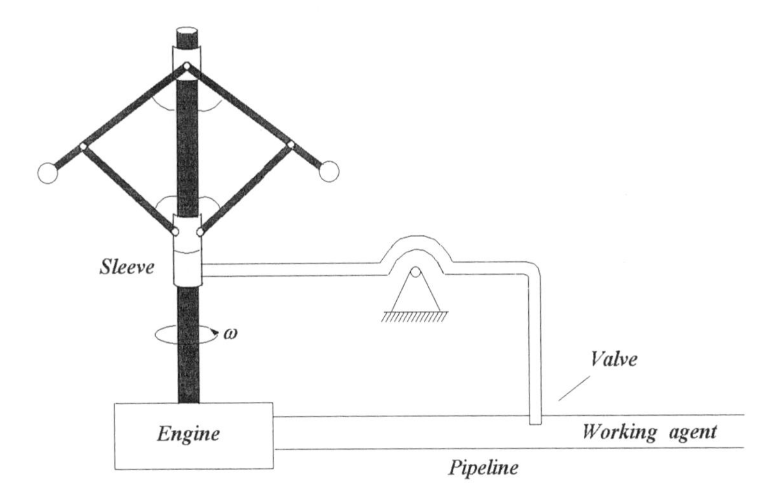

Fig. 1.1

In Fig. 1.1 is shown a basic diagram of such a governor. The two identical arms with equal governor flyballs at their ends are hung at the top of a shaft. The arms are fastened by additional hinges and can leave the vertical by the angle φ. The arms and shaft are in the same plane. Leaving the vertical by the angle φ, the arms set in motion the sleeve fitting on a shaft.

The working agent (steam, water, diesel fuel) is piped. The pipe has a butterfly valve. The working agent creates a rotational moment of a shaft, which the Watt governor is placed on. In the case of a steam turbine, for instance, the steam jet acts on turbine blades fitting on the shaft and creates a moment of force F. The equation, relating the moment of force F and the angular velocity of shaft $\omega(t)$, is as follows

$$J\frac{d\omega}{dt} = F - G. \tag{1.1}$$

Here J is a moment of inertia of rotating rigid body (in the case of a turbogenerator it consists of a shaft and a rotor of electric generator, which are rigidly connected to one another). For simplicity we neglect the masses of various parts of the Watt governor since they are negligible compared to J. The moment of force G is a sum of a useful load and a drag torque. At power stations, for example, G forms an electrical network power. Equation (1.1) is well known in theoretical mechanics. The determination of a moment of inertia J is an applied problem of calculus.

Recall that the Watt governor is employed for a given angular velocity to be kept constant, in which case $\omega(t) = \omega_0$. The value ω_0 is determined by specific requirements of a concrete technology problem. For a turbogenerator, for example, a very important condition is that ω_0 must coincide with the frequency of electric current.

The Watt governor operates in the following way. If the value of $\omega(t)$ is greater than ω_0, then the centrifugal force (and therefore the angle $\varphi(t)$) increases, the sleeve is raised and closes a valve by means of levers, reducing the working agent. This leads to reducing the angular velocity $\omega(t)$, and vice versa, if $\omega(t)$ decreases, then the centrifugal force (and therefore the angle $\varphi(t)$) also decreases, the sleeve has fallen, the valve is raised, and the flow of working agent increases.

Such a control method is called a *negative feedback.* If the controlled value is greater than a certain given value, then the governor acts in such a way that this value decreases and vice versa if the controlled value is less than the given one, then the governor acts in such a way that this value increases.

In Fig. 1.2 is shown a block diagram, which is often used for describing control processes.

Here the rectangles may be regarded as some operators, which take each element of one functional space to elements of the other functional space. The operator "engine" takes each element of $\{\varphi\}$ to elements of $\{\omega\}$ and the operator "the Watt governor" takes each element of $\{\omega\}$ to elements of $\{\varphi\}$. Therefore we regard functions $\varphi(t)$ as inputs of the block "engine", and functions $\omega(t)$ as outputs of this block. For the block "the Watt governor", $\omega(t)$ is an input and $\varphi(t)$ is an output. From equation (1.1) it follows that the input does not always determine the output uniquely. For this reason the "initial data" (in the case considered it is $\omega(0)$) must be given.

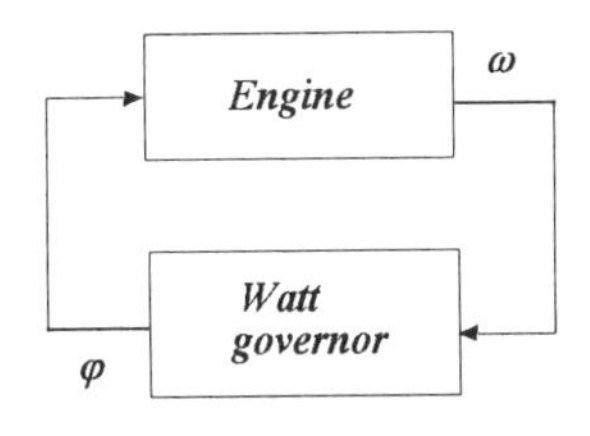

Fig. 1.2

In the second half of 19th century it has become obvious that the value $\omega(t)$ control cannot always be performed by means of the Watt governor. The development of technology has led to new types of steam engines, for which the Watt governors are practically useless.

In studying the problems, related with the Watt governor operation, the first scholars of the day, such as I.A.Vyshnegradsky, J.Maxwell, and A.Stodola arrived at the conclusion being unexpected for engineers and nontrivial for mathematicians.

The main ideas of these scholars are discussed below. For simplicity we consider a certain modification of the classic Watt governor rather than it himself.

Consider one of these modifications, namely a turbine shaft control system, which is often used in a turbine construction. In Fig. 1.3 is shown a schematic diagram of such a system.

The two equal governor flyballs of masses m, which are fitted on the bar, are fixed by springs and can slide along the bar. Here, unlike the previous case, a bar is used rather than hinges. The governor flyballs are also joined together by means of a membrane. The value of bending of the membrane depends on a motion of flyballs along the bar. The special oil fills a chamber (which is above the membrane) of device. This device is called a servomechanism. The oil fulfills here two functions. It lubricates all moving details and moves the piston, rigidly connected with the valve. A constant forcing pressure induces an inflow of oil, which is assumed to be an incompressible fluid. The quantity of an outflowing oil depends on the bending of membrane.

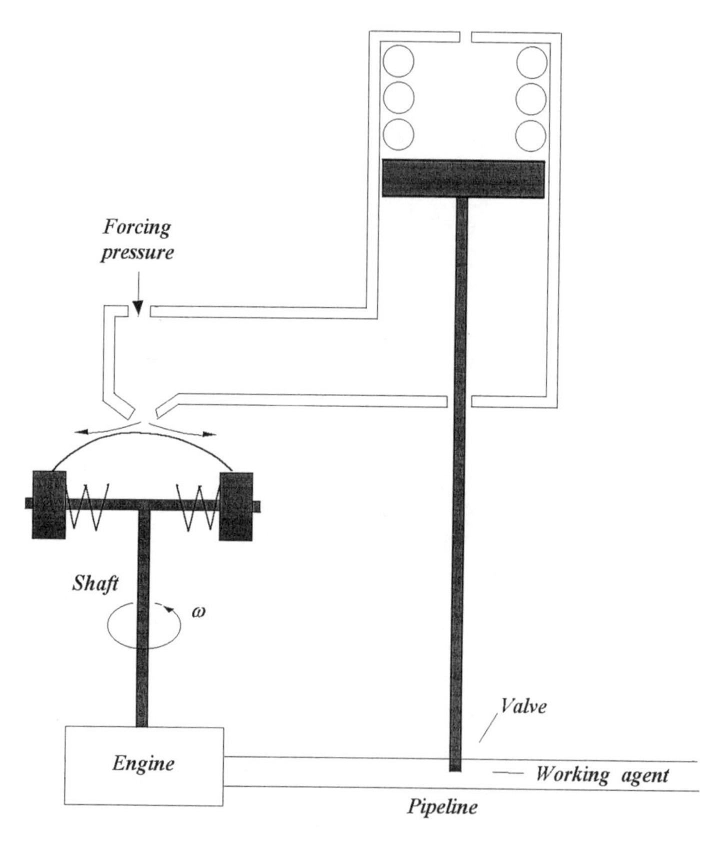

Fig. 1.3

Let us now consider a negative feedback. If the angular velocity of the shaft $\omega(t)$ decreases, then the centrifugal force, acting on the flyballs, also decreases and the springs tighten the flyballs to the center of shaft. In this case the bending of membrane increases and a size of the outlet decreases. As a result the amount of fluid, running out the outlet, decreases and its volume increases. The fluid, remaining in the chamber of servomechanism, moves up the piston, opening the valve and increasing thus the flow of working agent.

If the angular velocity $\omega(t)$ increases, then the valve comes down and cuts off an excess of working agent.

Beginning from 1952 such governors have been arranged on turbines at the Leningrad Metallic Factory [41].

Now we consider the second modification of the Watt governor. The functions of it are the same as before (see Fig. 1.1 and Fig. 1.3) but the mathematical description, from the methodological point of view, is much simpler than a description of the governors mentioned above.

Such a governor is shown in Fig. 1.4.

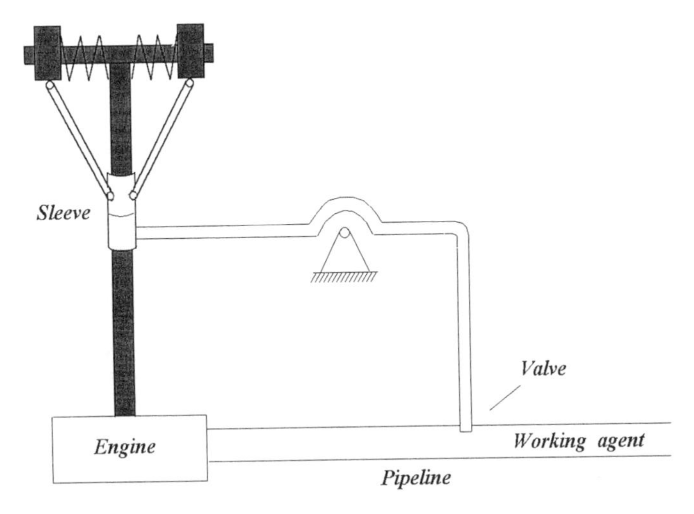

Fig. 1.4

The only difference between the operation of the classic Watt governor and the system, shown in Fig. 1.4, is as follows. In the case of the system mentioned above the centrifugal force f acts on the flyballs of masses m, which slide along a bar and are connected with the shaft by springs. At time t this force declines the masses from the stress-free position of spring by a value $x(t)$. Thus the output of the modified Watt governor is a value $x(t)$, which is transmitted into a device to move a valve by a distance u depending on x. Further we are interested in the fact of the existence of a functional dependence $u(x)$ only, leaving aside the design of this device.

Recall that $x(t)$ and f satisfy the following equations

$$f = \beta m(x + r)\omega^2, \tag{1.2}$$

$$m\ddot{x} + \alpha\dot{x} + \gamma x = f. \tag{1.3}$$

Here α, β, γ, r are positive numbers. The value $-\gamma x$ corresponds to an elastic force of the spring (the Hooke's law), the value $-\alpha\dot{x}$ corresponds to a friction force. We also assume that the viscous friction law is valid, i.e., the friction is proportional to the velocity $\dot{x}(t)$ and the number $-\alpha$ is a coefficient of proportionality. The other of the kinds of friction is neglected. The number r denotes the length of spring in the stress–free state.

Under the assumption that G is constant and F is a function of u, namely $F = F(u)$, by (1.1)–(1.3) we have the following system of equations for describing a value $\omega(t)$ control

$$\begin{aligned} J\dot{\omega} &= F(u(x)) - G, \\ m\ddot{x} + \alpha\dot{x} + \gamma x &= \beta m(x + r)\omega^2. \end{aligned} \tag{1.4}$$

We see that for the normal functioning of this system, the equations (1.4) must have a solution of the form

$$\omega(t) \equiv \omega_0, \qquad x(t) \equiv x_0. \tag{1.5}$$

Here ω_0 is a required angular velocity of shaft and x_0 is a number satisfying the following equations

$$\begin{aligned} F(u(x_0)) &= G, \\ \gamma x_0 &= \beta m(x_0 + r)\omega_0^2. \end{aligned} \tag{1.6}$$

It follows that equations (1.6) are necessary and sufficient conditions for the existence of the solution of system (1.4) in the form (1.5). Equations (1.6) can be obtained in the following way. From physical considerations we can conclude that the first equation of (1.6) has a root x_0. The second equation is satisfied by the choice of a value of spring deflection rate and by using x_0, obtained previously.

In this case a stability of trivial solution (1.5) is a main scientific and technical problem, a solving of which was far nontrivial and fully unexpected for engineers.

Consider now system (1.4). Let us linearize it in a neighborhood of solutions (1.5) under the assumption that relations (1.6) are satisfied and the function $F(u(x))$ is sufficiently smooth. Denoting $\Delta x(t) = x(t) - x_0$ and $\Delta\omega(t) = \omega(t) - \omega_0$, we obtain the following linear system

$$\begin{aligned} &J(\Delta\omega)^{\bullet} = F_0 \Delta x, \\ &m(\Delta x)^{\bullet\bullet} + \alpha(\Delta x)^{\bullet} + \gamma \Delta x = \beta m \omega_0^2 \Delta x + 2\beta m \omega_0 (x_0 + r)\Delta\omega. \end{aligned} \tag{1.7}$$

Here $F_0 = F'(u(x_0))u'(x_0)$, the higher order terms are omitted, and all the functions are replaced by their linear approximations.

The linearizations of the type (1.7) are usually called equations of linear approximation.

We now consider the equations of linear approximation (1.7) and then a relation between equations (1.7) and original equations (1.4).

System (1.7) is equivalent to the following equation of the third order

$$(\Delta\omega)^{\bullet\bullet\bullet} + \frac{\alpha}{m}(\Delta\omega)^{\bullet\bullet} + \frac{\gamma - \beta m \omega_0^2}{m}(\Delta\omega)^{\bullet} - \frac{f_0 F_0}{Jm}\Delta\omega = 0, \tag{1.8}$$

where $f_0 = 2\beta m \omega_0 (x_0 + r)$. The characteristic polynomial of such an equation has the form

$$Q(p) = p^3 + \frac{\alpha}{m}p^2 + \frac{\gamma - \beta m \omega_0^2}{m}p - \frac{f_0 F_0}{Jm}. \tag{1.9}$$

Consider equation (1.8). From the elementary theory of integration it follows that any solution $\Delta\omega(t)$, corresponding to small initial data $\Delta\omega(0)$, $(\Delta\omega(0))^{\bullet}$, $(\Delta\omega(0))^{\bullet\bullet}$, remains small and tends to zero as $t \to +\infty$ if and only if all zeros of the polynomial $Q(p)$ have negative real parts. In this case the solution of linear system is called asymptotically stable. A polynomial $Q(p)$, all zeros of which have negative real parts, is often called a stable polynomial.

1.2 The Hermite—Mikhailov criterion

Does it make possible to say anything about a stability of the polynomial $Q(p)$ without finding its zeros? (For mathematicians this question was raised in 60s of the 19th century in connection with the problem of studying the Watt governor). For a polynomial of the second degree $Q(p) = p^2 + \alpha p + \beta$ the answer to this question is sufficiently simple. This is a necessary and sufficient condition for α and β to be positive. In this case for a polynomial of the third degree of the type of (1.9) the answer to the question is considerably complicated.

At present there are many various criteria of a stability of polynomials, namely the criteria of Hurwitz, Routh, Lienard. These criteria can be found in the remarkable book [16]. A criterion being the most simple and popular among engineers is a criterion of Hermite-Mikhailov. It will be discussed below.

Consider a polynomial of degree n with the real coefficients

$$f(p) = p^n + a_{n-1}p^{n-1} + \ldots + a_0.$$

We state a simple fact, which is sometimes called the Stodola theorem.

Proposition 1.1. *If all zeros of the polynomial $f(p)$ have negative real parts, then all its coefficients a_i are positive.*

Proof. Denote by α_j real zeros of $f(p)$ and by β_k the complex ones. Since a_i are real, the numbers $\bar{\beta}_k$ are also zeros of $f(p)$.

Thus, the polynomial $f(p)$ may be represented in the form of the following product

$$f(p) = \prod_j (p - \alpha_j) \prod_k (p - \beta_k)(p - \bar{\beta}_k) =$$
$$= \prod_j (p - \alpha_j) \prod_k (p^2 - 2(\operatorname{Re}\beta_k)p + |\beta_k|^2).$$

If for all j and k the conditions $\alpha_j < 0$ and $\operatorname{Re}\beta_k < 0$ are satisfied, then the following products

$$\prod_j (p - \alpha_j), \qquad \prod_k (p^2 - 2(\operatorname{Re}\beta_k)p + |\beta_k|^2)$$

are polynomials with positive coefficients. Consequently, the polynomial $f(p)$ has also positive coefficients.

Denote by m the number of zeros of $f(p)$ with the positive real parts. Consider the values of polynomial $f(p)$ on the imaginary axis:

$$f(i\omega), \quad \omega \in \mathbb{R}^1$$

A curve on the complex plane, $\{p = f(i\omega) | \omega \in \mathbb{R}^1\}$, is called a hodograph of polynomial $f(p)$. Sometimes this hodograph is called the Mikhailov hodograph.

Consider a function

$$\varphi(\omega) = \operatorname{Arg} f(i\omega).$$

Here $\operatorname{Arg} z$ is a continuous branch of the many-valued function:

$$\arg z + 2k\pi, \tag{1.10}$$

where k is an integer and $\arg z$ is an argument of z such that $-\pi < \arg z \leq \pi$. For definiteness, we assume that $\varphi(0) = 0$. If the hodograph crosses a ray in the complex plane $\{\operatorname{Im} z = 0, \ \operatorname{Re} z \leq 0\}$ at the point $\omega = \omega_0$, then we take a branch of function (1.10), which assures the continuity of the function $\varphi(\omega)$ at the point ω_0.

Denote by $\Delta\varphi(\omega)\big|_{-\infty}^{+\infty}$ an increment of the function $\varphi(\omega)$ in the case that the argument ω runs from $-\infty$ to $+\infty$.

Proposition 1.2. *If $f(i\omega) \neq 0, \forall\, \omega \in \mathbb{R}^1$, then the formula of Hermite-Mikhailov*

$$\Delta\varphi(\omega)\big|_{-\infty}^{+\infty} = \pi(n - 2m)$$

is valid.

Proof. Denote by α_j zeros of $f(p)$ with the negative real parts and by β_k zeros of $f(p)$ with the positive real parts.

Under the above assumption the polynomial $f(p)$ does not have zeros on the imaginary axis.

Considering the polynomial $f(p)$ as a product

$$f(p) = \prod_j (p - \alpha_j) \prod_k (p - \beta_k)$$

and applying the well-known theorems on an argument of a product of complex numbers, we obtain

$$\Delta\varphi(\omega)\big|_{-\infty}^{+\infty} = \sum_j \Delta\operatorname{Arg}(i\omega - \alpha_j)\Big|_{-\infty}^{+\infty} + \sum_k \Delta\operatorname{Arg}(i\omega - \beta_k)\Big|_{-\infty}^{+\infty}. \tag{1.11}$$

Now let us compute the values

$$\Delta\operatorname{Arg}(i\omega - \alpha_j)\big|_{-\infty}^{+\infty}, \qquad \Delta\operatorname{Arg}(i\omega - \beta_k)\big|_{-\infty}^{+\infty}.$$

For this purpose we consider on the complex plane (Fig. 1.5) the numbers α_j, $i\omega$, $i\omega - \alpha_j$ and the corresponding to them vectors.

With increasing ω from $-\infty$ to $+\infty$ the end of vector, corresponding to the number $i\omega - \alpha_j$, monotonically slides along the line $\operatorname{Re} z = -\operatorname{Re}\alpha_j$ and rotates counterclockwise as it is shown in Fig. 1.5. Whence it follows that

$$\Delta \operatorname{Arg}\,(i\omega - \alpha_j)\Big|_{-\infty}^{+\infty} = \pi.$$

Consider now the numbers β_k, $i\omega$, $i\omega - \beta_k$ on the complex plane (Fig. 1.6) and the vectors, corresponding to these numbers.

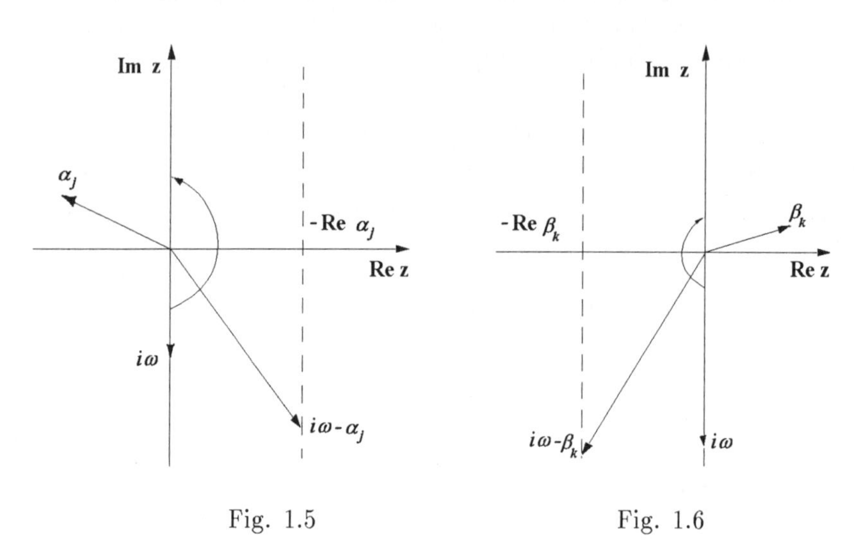

Fig. 1.5 Fig. 1.6

As ω increases from $-\infty$ to $+\infty$ the end of vector, corresponding to the number $i\omega - \beta_k$, monotonically slides along the line $\operatorname{Re} z = -\operatorname{Re}\beta_k$ and rotates clockwise as it is shown in Fig. 1.6. Consequently,

$$\Delta \operatorname{Arg}\,(i\omega - \beta_k)\Big|_{-\infty}^{+\infty} = -\pi.$$

From decomposition (1.11) it follows that

$$\Delta\varphi(\omega)\Big|_{-\infty}^{+\infty} = (n-m)\pi - m\pi = \pi(n-2m).$$

The proof of Proposition 1.2 is completed.

The following criterion results from Proposition 1.2.

The Hermite-Mikhailov criterion. *Let $f(i\omega) \neq 0$, $\forall\, \omega \in \mathbb{R}^1$. Then for a polynomial $f(p)$ to be stable it is necessary and sufficient that*

$$\Delta\varphi(\omega)\Big|_{-\infty}^{+\infty} = n\pi. \tag{1.12}$$

The following remark allows us to simplify a verification of the test (1.12). Since the coefficients of polynomial $f(p)$ are real, we have the following relations

$$\operatorname{Re} f(-i\omega) = \operatorname{Re} f(i\omega), \qquad \operatorname{Im} f(-i\omega) = -\operatorname{Im} f(i\omega).$$

This implies that the hodograph $f(p)$ is symmetric with respect to the real axis (see Fig. 1.7)

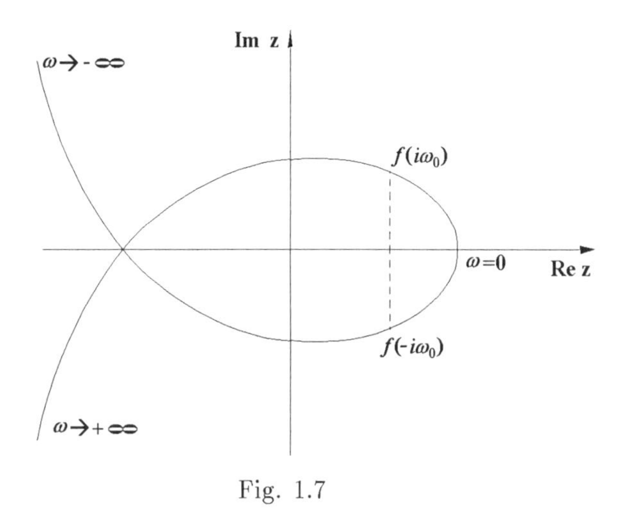

Fig. 1.7

Then in place of formula (1.12) it makes possible to consider the following condition

$$\Delta\varphi(\omega)\Big|_0^{+\infty} = \frac{n\pi}{2}.$$

Note that together with the proof of Proposition 1.2 we have also proved the fact that for stable polynomials the function $\varphi(\omega)$ grows monotonically. In this case with increasing ω the hodograph of $f(p)$ for $\omega \geq 0$, going out of the point $\omega = 0$ on the positive semi-axis $\{\operatorname{Re} z > 0,\ \operatorname{Im} z = 0\}$, sequentially crosses the semi-axes $\{\operatorname{Re} z = 0,\ \operatorname{Im} z > 0\}$, $\{\operatorname{Im} z = 0,\ \operatorname{Re} z < 0\}$,... while passing n quadrants. If $f(i\omega) \neq 0\ \forall \omega \in \mathbb{R}^n$ and with increasing ω the hodograph of $f(p)$ for $\omega \geq 0$, going out of a point on the semi-axis $\{\operatorname{Re} z > 0,\ \operatorname{Im} z = 0\}$, sequentially crosses the semi-axes $\{\operatorname{Re} z = 0, \operatorname{Im} z > 0\}$, $\{\operatorname{Re} z < 0,\ \operatorname{Im} z = 0\}$, ..., asymptotically tending to the $n-$th semi-axis in order, then $f(p)$ is a stable polynomial (by formula 1.12).

Consider a polynomial of the third degree

$$f(p) = p^3 + \alpha p^2 + \beta p + \gamma.$$

By Proposition 1.1, a necessary condition for $f(p)$ to be stable is that the conditions $\alpha > 0$, $\beta > 0$, $\gamma > 0$ be satisfied.

Here

$$f(i\omega) = (\gamma - \alpha\omega^2) + (\beta - \omega^2)\omega i.$$

Then the crosspoints of the hodograph of $f(i\omega)$ for $\omega \geq 0$ with the semi-axes $\{\operatorname{Re} z > 0,\ \operatorname{Im} z = 0\}$, $\{\operatorname{Re} z = 0,\ \operatorname{Im} z > 0\}$, $\{\operatorname{Re} z < 0,\ \operatorname{Im} z = 0\}$ correspond to the following points

$$\omega_1 = 0, \quad \omega_2 = \sqrt{\gamma/\alpha}, \quad \omega_3 = \sqrt{\beta},$$

In this case $f(0) = \gamma$, $f(i\omega_2) = (\beta - \gamma/\alpha)\sqrt{\gamma/\alpha}\, i$, $f(i\omega_3) = \gamma - \alpha\beta$.

Whence it follows that a polynomial is stable if and only if

$$\beta - \frac{\gamma}{\alpha} > 0, \qquad \gamma - \alpha\beta < 0. \tag{1.14}$$

In addition, we have

$$\frac{\operatorname{Re} f(i\omega)}{\operatorname{Im} f(i\omega)} \to 0 \tag{1.15}$$

as $\omega \to +\infty$. Thus, if (1.14) and (1.15) are satisfied, then with increasing ω the hodograph of $f(p)$ for $\omega \geq 0$, going out of a point on the semi-axis $\{\operatorname{Re} z > 0,\ \operatorname{Im} z = 0\}$, crosses the semi-axes $\{\operatorname{Re} z = 0,\ \operatorname{Im} z > 0\}$, $\{\operatorname{Re} z < 0,\ \operatorname{Im} z = 0\}$ sequentially and exactly one time and asymptotically tends to the third semi-axis $\{\operatorname{Re} z = 0,\ \operatorname{Im} z < 0\}$ and vice versa such a behavior of the hodograph is possible only if (1.14) is satisfied.

Thus, for the polynomial to be stable it is necessary and sufficient for the following inequalities

$$\alpha > 0, \quad \beta > 0, \quad \gamma > 0, \quad \alpha\beta > \gamma \tag{1.16}$$

to be satisfied. Sometimes these inequalities are called the condition of Vyshnegradsky.

1.3 Theorem on stability by the linear approximation

Consider a solvability of a matrix equation

$$A^*H + HA = G, \tag{1.17}$$

which is often called the Lyapunov equation. Here A and G are the given $n \times n$-matrices and $n \times n$-matrix H is a solution of (1.17). We consider the case that H and G are symmetric.

Lemma 1.1. *Suppose, all the eigenvalues of matrix A have negative real parts and $G < 0$. Then equation* (1.17) *has a unique solution*

$$H = -\int_0^{+\infty} e^{A^*t} G e^{At} dt. \tag{1.18}$$

Recall that the inequality $G < 0$ means that the corresponding quadratic form z^*Gz is negative definite. We also note that from (1.18) it follows that $H > 0$. Really, the nondegeneracy of a matrix e^{At} implies that

$$-(e^{At})^* G e^{At} > 0, \quad \forall t \geq 0.$$

Hence for any $x \in \mathbb{R}^n$ we have

$$-x^*(e^{At})^* G e^{At} x > 0, \qquad \forall t \geq 0.$$

Then

$$x^*Hx = -\int_0^{-\infty} x^*(e^{At})^* G e^{At} x > 0.$$

P r o o f of L e m m a 1.1. The assumptions on the eigenvalues of the matrix A imply that finite integral (1.18) exists.

It is obvious that

$$\frac{d}{dt}(e^{A^*t} G e^{At}) = A^*(e^{A^*t} G e^{At}) + (e^{A^*t} G e^{At})A.$$

Integrating this identity from 0 to $+\infty$ and taking into account that

$$\lim_{t \to +\infty} e^{A^*t} G e^{At} = 0,$$

we obtain equation (1.17) with H, satisfying (1.18).

Now we show the uniqueness of the solution of equation (1.17). Suppose the contrary, i.e., that H_1 and H_2 are two solutions of equation (1.17). Then $H = H_1 - H_2$ satisfies an equation

$$A^*H + HA = 0. \tag{1.19}$$

Consider a vector function

$$x(t) = e^{At}x_0,$$

where x_0 is a vector.

By (1.19) we have

$$\frac{d}{dt}\left(x(t)^*Hx(t)\right) = x(t)^*(A^*H + HA)x(t) \equiv 0.$$

Hence,

$$x(t)^*Hx(t) \equiv x_0^*Hx_0.$$

However $x(t) \to 0$ as $t \to +\infty$ by virtue of the assumptions on eigenvalues of A.

Thus, we have $x_0^*Hx_0 = 0$, $\forall\, x_0 \in \mathbb{R}^n$. Whence it follows that the symmetric matrix H is null.

This completes the proof of lemma.

Consider now a differential equation

$$\frac{dx}{dt} = f(t,x), \qquad t \in \mathbb{R}^1, \quad x \in \mathbb{R}^n, \tag{1.20}$$

where $f(t,x)$ is a continuous vector function: $\mathbb{R}^1 \times \mathbb{R}^n \to \mathbb{R}^n$. Further we assume that all the solutions $x(t,t_0,x_0)$ with initial data $x(t_0,t_0,x_0) = x_0$ are defined on the interval $(t_0,+\infty)$.

Definition 1.1. *A solution $x(t,t_0,x_0)$ of system (1.20) is said to be Lyapunov stable if for any number $\varepsilon > 0$ there exists a number $\delta(\varepsilon) > 0$ such that for all y_0, satisfying the inequality $|x_0 - y_0| \le \delta(\varepsilon)$, the following relation holds*

$$\left|x(t,t_0,x_0) - x(t,t_0,y_0)\right| \le \varepsilon, \qquad \forall\, t \ge t_0. \tag{1.21}$$

Definition 1.2. *If a solution $x(t,t_0,x_0)$ is Lyapunov stable and there exists a number δ_0 such that for all y_0, satisfying the inequality $|x_0 - y_0| \leq \delta_0$, the following equation*

$$\lim_{t\to+\infty} \left|x(t,t_0,x_0) - x(t,t_0,y_0)\right| = 0$$

is valid, then the solution $x(t,t_0,x_0)$ is called asymptotically stable.

Note that, generally speaking, in Definitions 1.1 and 1.2 the numbers $\delta(\varepsilon)$ and δ_0 also depend on t_0: $\delta_0 = \delta_0(t_0)$, $\delta(\varepsilon) = \delta(\varepsilon, t_0)$. If δ_0 and $\delta(\varepsilon)$ can be chosen independent of t_0, then the solution $x(t,t_0,x_0)$ is called uniformly Lyapunov stable and uniformly asymptotically stable.

The Lyapunov instability is a logical negation of the Lyapunov stability.

Below the examples of stable and unstable solutions are given.

Consider the equation of pendulum

$$\ddot{\theta} + \alpha\dot{\theta} + (g/l)\sin\theta = 0. \tag{1.22}$$

Here $\theta(t)$ is an angle of deviation of the pendulum from the vertical position, l is a length (Fig. 1.8), g is acceleration of gravity and α is a friction factor.

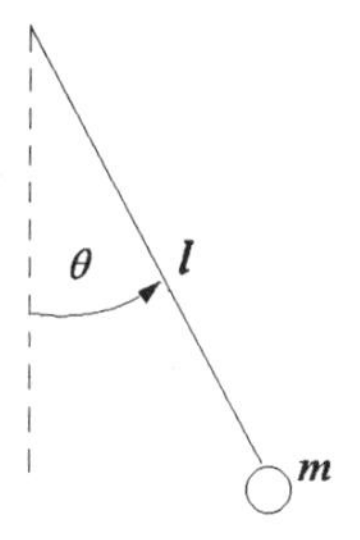

Fig. 1.8

Equation (1.22) can be given in the form (1.20):

$$\begin{aligned} \dot{\theta} &= \eta, \\ \dot{\eta} &= -\alpha\eta - \frac{g}{l}\sin\theta. \end{aligned} \tag{1.23}$$

Schematic representation of the two-dimensional phase space, filled by trajectories of system (1.23), which is often called a phase portrait, is shown for the case $\alpha = 0$ in Fig. 1.9 and that for the case $\alpha > 0$ in Fig. 1.10. For

$\alpha = 0$ system (1.23) has the following first integral

$$V(\theta, \eta) = \eta^2 - \frac{2g}{l} \cos\theta = C, \tag{1.24}$$

where C is an arbitrary number.

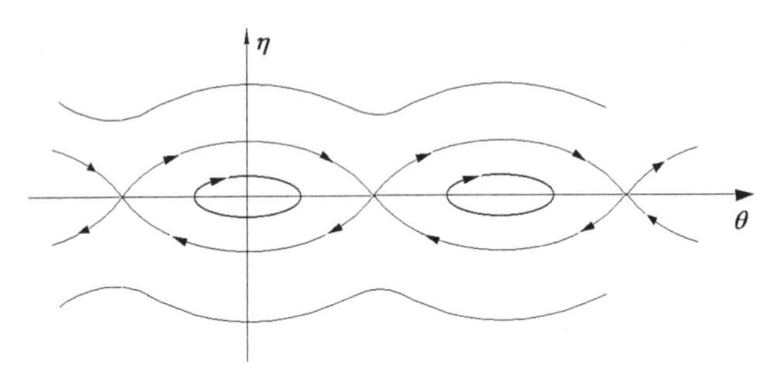

Fig. 1.9

We see that any solution $\theta(t), \eta(t)$ of system (1.23) satisfies the following identity

$$\frac{d}{dt} V(\theta(t), \eta(t)) = 2\eta(t) \left(-\frac{g}{l} \sin\theta(t)\right) + \frac{2g}{l} (\sin\theta(t))\, \eta(t) \equiv 0.$$

Thus, the trajectories in the phase space of system (1.23) are placed wholly on the level lines

$$\{\theta, \eta |\ V(\theta, \eta) = C\}.$$

Whence it follows that trajectories are closed in the neighborhood of a stationary solution $\theta(t) \equiv 0$. This implies that the solution is Lyapunov stable but not asymptotically stable.

Using the first integral (1.24) we see that two trajectories tend to stationary solution $\theta(t) \equiv \pi, \eta(t) \equiv 0$ as $t \to -\infty$ and the same trajectories tend to equilibria $\theta(t) \equiv -\pi,\ \eta(t) \equiv 0$ and $\theta(t) \equiv 3\pi,\ \eta(t) \equiv 0$ as $t \to +\infty$. Such trajectories are often called heteroclinic. Their existence proves that the solution $\theta(t) \equiv \pi, \eta(t) \equiv 0$ is Lyapunov unstable.

The first stationary solution, which is Lyapunov stable, corresponds to the lower equilibrium position of the pendulum. In some neighborhood of this position the closed trajectories correspond to the periodic oscillations of a pendulum in a neighborhood of the lower equilibrium position.

The second stationary solution, which is Lyapunov unstable. corresponds to the upper equilibrium position. The latter exists theoretically but we cannot observe it because of its instability. This fact is the same for many other physical, technical, biological, and economical systems, that is, the Lyapunov unstable equilibrium is nonrealizable.

We consider an intuitive "mechanical" proof of asymptotic stability of the lower equilibrium position $\theta(t) \equiv 0, \eta(t) \equiv 0$ for $\alpha > 0$ only. If $\alpha > 0$, then there exist friction forces, which assure the decay of oscillation near the lower equilibrium position. Thus, solutions in a neighborhood of the stationary point $\theta(t) \equiv 0, \eta(t) \equiv 0$ tend to zero as $t \to +\infty$. It means that the stationary solution is asymptotically stable (Fig. 1.10).

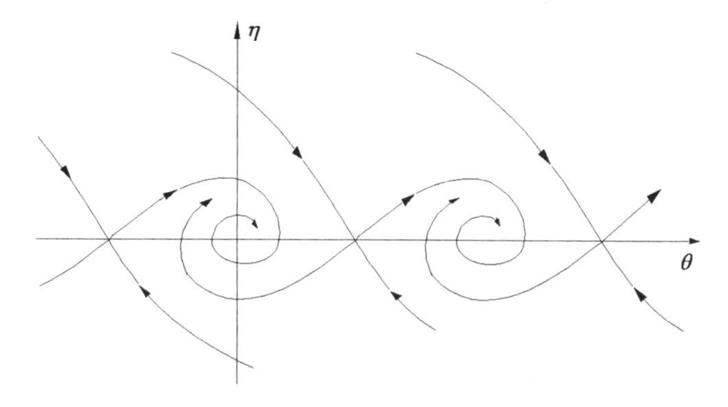

Fig. 1.10

A.M. Lyapunov has suggested a method of investigation of a solution stability, involving special functions, which are called now the Lyapunov functions. Consider the case that the solution $x(t, t_0, x_0)$ is the zero solution, namely $x(t, t_0, x_0) \equiv 0$. The general case may be reduced to it by the change of variable

$$x = y + x(t, t_0, x_0).$$

Then

$$\frac{dy}{dt} = g(t, y), \tag{1.25}$$

where $g(t, y) = f(t, y + x(t, t_0, x_0)) - f(t, x(t, t_0, x_0))$.

We see that equation (1.25) has the same structure as (1.20) and in addition we have $g(t, 0) \equiv 0$. However such a substitution is not always effective since in this case we must know the form of solution $x(t, t_0, x_0)$.

Consider a differentiable in some neighbourhood of the point $x = 0$ function $V(x)$ $(V : \mathbb{R}^n \to \mathbb{R}^1)$ such that $V(0) = 0$. We introduce the following notation

$$\dot{V}(x) := (\operatorname{grad} V(x))^* f(t,x) = \sum_i \frac{\partial V}{\partial x_i} f_i(t,x).$$

The expression $\dot{V}(x)$ is often called a derivative of the function $V(x)$ with respect to system (1.20). Here x_i is the i–th component of the vector x and f_i is the i–th component of the vector function f. It is clear that if we take the solution $x(t,t_0,x_0)$ rather than x, then by the differentiation rule the following relation holds

$$\frac{d}{dt} V(x(t,t_0,x_0)) = \big(\operatorname{grad} V(x(t,t_0,x_0))\big)^* f\big(t, x(t,t_0,x_0)\big).$$

Theorem 1.1 (On asymptotic stability). *Suppose, there exist a differentiable function $V(x)$ and a continuous function $W(x)$ such that in some neighborhood of the point $x = 0$ the following conditions hold:*

1) $V(x) > 0$ *for* $x \neq 0$, $V(0) = 0$,
2) $\dot{V}(x) \leq W(x) < 0$ *for* $x \neq 0$.

Then the zero solution of system (1.20) *is asymptotically stable.*

Proof. Assume that the ball $\{x|\ |x| \leq \varepsilon\}$ is inside the neighborhood considered. Put

$$\alpha = \inf_{\{x|\ |x|=\varepsilon\}} V(x). \tag{1.26}$$

Since the sphere is closed, assumption 1) implies that $\alpha > 0$.

Let us choose now a number δ such that

$$\sup_{\{x|\ |x|\leq\delta\}} V(x) < \alpha. \tag{1.27}$$

The existence of such a δ follows from the relation $V(0) = 0$ and a continuity of function $V(x)$. We shall prove that for initial data x_0 such that $|x_0| \leq \delta$ the inequality $|x(t,t_0,x_0)| \leq \varepsilon$, $\forall t \geq t_0$ is satisfied, i.e., the Lyapunov stability is valid.

Suppose the contrary. Then by the continuity of the solution $x(t,t_0,x_0)$ there exists a number $\tau > t_0$ such that $|x(\tau,t_0,x_0)| = \varepsilon$ and

$$\big|x(t,t_0,x_0)\big| \leq \varepsilon, \qquad \forall\, t \in [t_0, \tau].$$

Under the assumption 2) we obtain the following inequality

$$V\big(x(\tau, t_0, x_0)\big) \leq V(x_0). \tag{1.28}$$

On the other hand, by (1.26) and (1.27)

$$V(x(\tau, t_0, x_0)) \geq \alpha > V(x_0). \tag{1.29}$$

Since (1.28) and (1.29) are in contrast to each other, we have $|x(t, t_0, x_0)| \leq \varepsilon$, $\forall t \geq t_0$.

Let us now prove the asymptotic stability mentioned above.

Holding a number ε_0 fixed such that a ball $\{x|\ |x| \leq \varepsilon_0\}$ is placed wholly inside the neighborhood of the point $x = 0$, we choose δ_0 such that $|x(t, t_0, x_0)| \leq \varepsilon_0, \ \ \forall t \geq t_0, \ \ \forall x_0 \in \{x|\ |x| \leq \delta_0\}$.

In this case assumption 2) yields that for any x_0 from the ball $\{x|\ |x| \leq \delta_0\}$ there exists a limit

$$\lim_{t \to +\infty} V(x(t, t_0, x_0)) = \beta \tag{1.30}$$

and $V(x(t, t_0, x_0)) \geq \beta$, $\forall t \geq t_0$.

This implies that $\beta = 0$. Assuming the opposite, that is, that $\beta > 0$, we obtain that the solution $x(t, t_0, x_0)$ is separated from zero, i.e., there exists a number γ such that

$$|x(t, t_0, x_0)| \geq \gamma, \qquad \forall t \geq t_0. \tag{1.31}$$

Recall that in addition to (1.31) the following condition

$$|x(t, t_0, x_0)| \leq \varepsilon_0, \qquad \forall t \geq t_0 \tag{1.32}$$

is also satisfied. Inequalities (1.31), (1.32), continuity of the function $W(x)$, and the inequality $W(x) < 0$, $\forall x \in \{x|\ \gamma \leq |x| \leq \varepsilon_0\}$ imply the existence of a negative number æ such that

$$W(x(t, t_0, x_0)) \leq æ.$$

It follows that

$$V(x(t, t_0, x_0)) \leq V(x_0) + \int_{t_0}^{t} W(x(\tau, t_0, x_0))\, d\tau \leq$$
$$\leq V(x_0) + æ(t - t_0) \underset{t \to +\infty}{\to} -\infty.$$

The last inequality contradicts the inequality $V(x(t,t_0,x_0)) \geq \beta > 0$. Thus, we have $\beta = 0$. From relation (1.30) and the continuity of $V(x)$ we conclude that

$$\lim_{t\to+\infty} |x(t,t_0,x_0)| = 0.$$

That establishes the theorem.

Theorem 1.2 (On instability). *Suppose that there exist a differentiable function $V(x)$ and a continuous function $W(x)$, for which in some neighborhood of the point $x = 0$ the following conditions hold:*

1) $V(0) = 0$ and for a sequence $x_k \to 0$ as $k \to \infty$ the inequalities $V(x_k) < 0$ are valid,

2) $\dot V(x) \leq W(x) < 0$ for $x \neq 0$.

Then the zero solution of system (1.20) is Lyapunov unstable.

Proof. Suppose the opposite, i.e., for $\varepsilon > 0$ a number $\delta(\varepsilon)$ can be found such that

$$|x(t,t_0,x_0)| \leq \varepsilon, \qquad \forall\, t \geq t_0$$

for all $x_0 \in \{x|\ |x| \leq \delta(\varepsilon)\}$. In this case by assumption 1) of the theorem we can choose x_0 such that $V(x_0) < 0$. Then from assumption 2) it follows that

$$V(x(t,t_0,x_0)) \leq V(x_0) < 0$$

and, consequently, there exists a number $\gamma > 0$ such that

$$|x(t,t_0,x_0)| \geq \gamma, \qquad \forall\, t \geq t_0.$$

Since $W(x)$ is continuous, a negative number æ can be found such that $W(x) \leq$ æ, $\forall\, x \in \{x|\ \gamma \leq |x| \leq \varepsilon\}$. Therefore

$$W(x(t,t_0,x_0)) \leq \text{æ}, \qquad \forall\, t \geq t_0.$$

Hence

$$V(x(t,t_0,x_0)) \leq V(x_0) + \int_{t_0}^{t} W(x(\tau,t_0,x_0))\,d\tau \leq$$
$$\leq V(x_0) + \text{æ}(t-t_0) \underset{t\to+\infty}{\to} -\infty,$$

which contradicts the assumption on the Lyapunov stability. The theorem is proved.

Consider now system (1.20) represented in the following form

$$\frac{dx}{dt} = Ax + g(t, x). \tag{1.33}$$

Here A is a constant $n \times n$-matrix, $g(t, x)$ is a continuous vector-function: $\mathbb{R}^1 \times \mathbb{R}^n \to \mathbb{R}^1$. Suppose that in some neighborhood of the point $x = 0$ the following inequality holds

$$|g(t, x)| \leq æ|x|, \qquad \forall t \in \mathbb{R}^1. \tag{1.34}$$

Here æ is some number. Assume that A does not have pure imaginary eigenvalues.

Let us construct functions $V(x)$ and $W(x)$, which satisfy the assumptions of Theorem 1.1 or Theorem 1.2.

At first, consider the case that all the eigenvalues of A have negative real parts. In this case for $G = -I$ by Lemma 1.1 a matrix $H > 0$ can be found such that

$$A^* H + HA = -I. \tag{1.35}$$

Let us consider further the quadratic form $V(x) = x^* Hx$, which is positive definite:

$$V(x) = x^* Hx > 0, \qquad \forall\, x \neq 0.$$

We observe that $V(x)$ satisfies assumption 1) of the theorem on asymptotic stability. Equation (1.35) can be rewritten as $2x^* HAx = -|x|^2$. Therefore, taking into account (1.33) and (1.34), we obtain

$$\dot{V}(x) = 2x^* H(Ax + g(t, x)) \leq -|x|^2 + 2|x^* H|\,|x|æ.$$

If æ satisfies an inequality

$$æ < (4|H|)^{-1}, \tag{1.36}$$

then assumption 2) of the theorem on asymptotic stability is also satisfied with $W(x) = -|x|^2/2$.

Thus, the following result can be stated.

Corollary 1.1. *If A is a stable matrix, i.e., all its eigenvalues have negative real parts and condition* (1.36) *holds, then the zero solution of system* (1.33) *is asymptotically stable.*

Consider the case that the matrix A has no pure imaginary eigenvalues and m of its eigenvalues have positive real parts. Without loss of generality we may assume that the matrix A has the following block representation

$$A = \begin{pmatrix} A_1 & 0 \\ 0 & -A_2 \end{pmatrix},$$

where A_1 is a stable $(n-m) \times (n-m)$-matrix, A_2 is a stable $m \times m$-matrix.

Applying again Lemma 1.1, we can prove the existence of symmetric matrices $H_1 > 0$ and $H_2 > 0$ of dimension $(n-m) \times (n-m)$ and $m \times m$ respectively such that the following relations hold

$$\begin{aligned} A_1^* H_1 + H_1 A_1 &= -I, \\ A_2^* H_2 + H_2 A_2 &= -I. \end{aligned} \tag{1.37}$$

A function $V(x) = x_1^* H_1 x_1 - x_2^* H_2 x_2$, satisfies assumption 1) of the instability theorem. Here

$$x = \begin{pmatrix} x_1 \\ x_2 \end{pmatrix}, \qquad x_1 \in \mathbb{R}^{n-m}, \quad x_2 \in \mathbb{R}^m.$$

Hence, using (1.37), we obtain

$$\dot{V}(x) = -|x|^2 + 2x_1^* H_1 g_1(t,x) - 2x_2^* H_2 g_2(t,x), \tag{1.38}$$

where $g_1(t,x)$ and $g_2(t,x)$ satisfy the relation

$$g(t,x) = \begin{pmatrix} g_1(t,x) \\ g_2(t,x) \end{pmatrix}.$$

Here $g_1(t,x) : \mathbb{R}^1 \times \mathbb{R}^n \to \mathbb{R}^{n-m}$, $g_2(t,x) : \mathbb{R}^1 \times \mathbb{R}^n \to \mathbb{R}^m$.

From (1.38) it follows that

$$\begin{aligned} \dot{V} &\le -|x|^2 + 2|H_1 x_1|\,|g_1(t,x)| + 2|H_2 x_2|\,|g_2(t,x)| \le \\ &\le -|x|^2 + 2(|H_1 x_1| + |H_2 x_2|)\,|g(t,x)| \le \\ &\le -|x|^2 + 2(|H_1 x_1| + |H_2 x_2|)\,æ|x|. \end{aligned}$$

Then for

$$æ < (4(|H_1| + |H_2|))^{-1} \tag{1.39}$$

assumption 2) of the instability theorem is satisfied with $W(x) = -|x|^2/2$.

We can state the following

Corollary 1.2. *If A does not have eigenvalues on the imaginary axis and is unstable (i.e., it has also eigenvalues with positive real parts), then under the assumption that inequality* (1.39) *is valid the zero solution is Lyapunov unstable.*

For the following autonomous system

$$\frac{dx}{dt} = f(x), \qquad x \in \mathbb{R}^n, \quad f(0) = 0 \tag{1.40}$$

with continuously differentiable vector function $f(x)$, Corollaries 1.1 and 1.2 may be stated in terms of the Jacobi matrix

$$\frac{\partial f(x)}{\partial x} = \begin{pmatrix} \frac{\partial f_1}{\partial x_1} & \cdots & \frac{\partial f_1}{\partial x_n} \\ \vdots & \vdots & \vdots \\ \frac{\partial f_n}{\partial x_1} & \cdots & \frac{\partial f_n}{\partial x_n} \end{pmatrix},$$

which is given at the point $x = 0$:

$$A = \left. \frac{\partial f}{\partial x} \right|_{x=0}.$$

In particular, the following result is true.

Theorem 1.3 (On stability by the linear approximation). *Let A have no pure imaginary eigenvalues.*

If A is stable, then the zero solution of system (1.40) *is asymptotically stable. If A is unstable, then the zero solution of system* (1.40) *is Lyapunov unstable.*

As an example, consider pendulum equation (1.23). At the point $\theta = 0, \eta = 0$ the Jacobi matrix is as follows

$$A = \begin{pmatrix} 0 & 1 \\ -g/l & -\alpha \end{pmatrix}.$$

Its characteristic polynomial takes the form

$$p^2 + \alpha p + \frac{g}{l}.$$

It is clear that for $\alpha > 0$ the matrix A is stable, i.e., both eigenvalues have negative real parts. Consequently, the equilibrium considered is asymptotically stable.

Now we consider the equilibrium $\theta = \pi, \eta = 0$. To apply the theorem, make the change of variables $\theta = \widetilde{\theta} + \pi$, $\eta = \widetilde{\eta}$. Then we obtain the system

$$\begin{aligned} \dot{\widetilde{\theta}} &= \widetilde{\eta}, \\ \dot{\widetilde{\eta}} &= -\alpha\widetilde{\eta} - \frac{g}{l}\sin(\widetilde{\theta} + \pi), \end{aligned}$$

whose the Jacobi matrix at the point $\widetilde{\theta} = 0, \widetilde{\eta} = 0$ takes the form

$$A = \begin{pmatrix} 0 & 1 \\ g/l & -\alpha \end{pmatrix}.$$

Its characteristic polynomial is given by

$$p^2 + \alpha p - \frac{g}{l}.$$

Obviously, for $\alpha \geq 0$ one of eigenvalues of A is positive and the other is negative. By the theorem, the solution is Lyapunov unstable.

We again consider the Watt governor. The characteristic polynomial of the Jacobi matrix of system (1.4) has the form (1.9). By Theorem 1.3 and condition (1.16) we conclude that stationary solution (1.5) is asymptotically stable if the following inequalities $F_0 < 0$,

$$\alpha(\gamma - \beta m\omega_0^2)J > -F_0 f_0 m \tag{1.41}$$

are satisfied and stationary solution (1.5) is Lyapunov unstable in the case that

$$\alpha(\gamma - \beta m\omega_0^2)J < -F_0 f_0 m. \tag{1.42}$$

This conclusion, which has been made by I.A.Vyshnegradsky in 1876, has impressed on his contemporaries. In the cases that the friction is small and (1.42) is satisfied, there occurs an effect being compared with the nonrealizability of the upper position of pendulum, in which case the required operating regime becomes nonrealizable due to its instability.

For the conclusion to be cogitable for engineers, I.A.Vyshnegradsky has stated his notorious "thesis":

The friction is a governing characteristic of a sensitive and correctly operating governor or shortly: "there is no governor without friction".

In the middle of 19th century an unstable operating regime of governors was explained by the fact that with increasing a machine power the more heave valves were used and for their control the greater masses of balls, m, were necessary. In this case the improvement of a surface treatment led to the considerable decrease of friction factor α. In addition for the working speed of machines to be increased it was necessary to decrease the moments of inertia J of a shaft and the connected with it details. Notice that since in the modern turbogenerators the value of J is large, inequalities (1.41) are always satisfied.

Thus, we obtain the conditions, which ensure the operation of system: a machine - the Watt governor. However in starting the system we must every time make a transition to the desired process from a given state of system. Such processes are called transient processes.

1.4 The Watt governor transient processes

Consider transient process for the Watt governor, using main conceptions of the method of Lyapunov functions. Suppose that the function $F(u(x))$ is linear

$$F(u(x)) - G = F_0 \Delta x = F_0(x - x_0)$$

(see equations (1.4)-(1.7)). This assumption is natural in the case that a spring constant γ is sufficiently large and the changing of $x(t)$ is sufficiently small. By the same argument the following approximation $f = \beta m r \omega^2 + \beta m \omega_0^2 x$ is used.

Thus, we have

$$\begin{aligned} &J\dot{\omega} = F_0 \Delta x, \\ &m(\Delta x)^{\bullet\bullet} + \alpha(\Delta x)^{\bullet} + \gamma_0(\Delta x) = \beta m r \omega^2 - \gamma_0 x_0 \end{aligned} \tag{1.43}$$

with initial data $\omega(0) = 0$, $\Delta x(0) = -x_0$, $(\Delta x(0))^{\bullet} = 0$, which correspond to an activation of system at time $t = 0$. Here $\gamma_0 = \gamma - \beta m \omega_0^2$.

We introduce the following notation

$$y = \frac{F_0}{J} \Delta x, \qquad z = \frac{F_0}{J} (\Delta x)^{\bullet}, \qquad a = \frac{\alpha}{m}, \qquad b = \frac{\gamma_0}{m},$$

$$\varphi(\omega) = \frac{(-F_0)}{mJ}\,(\beta m r \omega^2 - \gamma_0 x_0).$$

Equations (1.43) in these notation are given by

$$\begin{aligned} \dot{\omega} &= y, \\ \dot{y} &= z, \\ \dot{z} &= -az - by - \varphi(\omega). \end{aligned} \tag{1.44}$$

We shall use further a function

$$V(\omega, y, z) = a\int_0^{\omega} \varphi(x)\,dx + \varphi(\omega)y + \frac{by^2}{2} + \frac{(z+ay)^2}{2}.$$

This function has properties similar to those of functions $V(x)$ in the Lyapunov theorems on asymptotic stability and instability. Therefore the function $V(\omega, y, z)$ can also be called the Lyapunov function.

It is obvious that for any solution $\omega(t), y(t), z(t)$ of system (1.44) the following equation holds

$$\dot{V}(\omega(t), y(t), z(t)) = (\varphi'(\omega(t)) - ab)y(t)^2. \tag{1.45}$$

Lemma 1.2. *Let an inequality*

$$m\gamma_0 > \alpha^2 \tag{1.46}$$

be satisfied and on the interval $[0, \omega_1]$, where ω_1 is defined by

$$\int_0^{\omega_1} \varphi(x)\,dx = 0,$$

the following condition

$$ab > \varphi'(\omega) \tag{1.47}$$

be valid. Then for a solution of system (1.44) with initial data $\omega(0) = 0$, $y(0) = -\frac{(F_0)}{J}\,x_0$, $z(0) = 0$ the following inclusion holds

$$\omega(t) \in [0, \omega_1], \qquad \forall\, t \geq 0. \tag{1.48}$$

Proof. Obviously, for $t = 0$ inclusion (1.48) is true. We assume that for some $t > 0$ inclusion (1.48) is not valid. Then the continuous differentiability of the function $\omega(t)$ implies the existence of a number $\tau \geq 0$ such that one of the following relations holds:

1) $\omega(\tau) = 0, \ \ \dot\omega(\tau) \leq 0, \ \ \omega(t) \in [0, \omega_1], \ \ \forall t \in [0, \tau]$,

2) $\omega(\tau) = \omega_1, \ \ \dot\omega(\tau) \geq 0, \ \ \omega(t) \in [0, \omega_1], \ \ \forall t \in [0, \tau]$.

Let us remark that in each of this cases by (1.47) we have the inequality $\dot V(\omega(t), y(t), z(t)) \leq 0, \forall t \in [0, \tau]$. From the relations

$$V(\omega(0), y(0), z(0)) = -\frac{\gamma_0}{2m}\frac{F_0^2}{J^2}x_0^2 + \frac{\alpha^2}{2m^2}\frac{F_0^2}{J^2}x_0^2 < 0$$

it follows that

$$V(\omega(\tau), y(\tau), z(\tau)) < 0. \tag{1.49}$$

In case 1) we have $\varphi(\omega(t))\, y(t) \geq 0$ since $\dot\omega(t) = y(t)$ and, consequently, $V(\omega(\tau), y(\tau), z(\tau)) \geq 0$. The last inequality contradicts inequality (1.49).

In case 2) we have $\varphi(\omega(t))\, y(t) \geq 0$ since $\omega_1 > \omega_0$ and therefore $V(\omega(\tau), y(\tau), z(\tau)) \geq 0$, which contradicts inequality (1.49).

These contradictions prove the lemma.

Lemma 1.3. *Let for a continuously differentiable on* $[0, +\infty)$ *function* $u(t)$ *the following assumptions hold:*

1) *for a number* C

$$|\dot u(t)| \leq C, \qquad \forall\, t \geq 0,$$

2) $u(t) \geq 0, \ \forall t \geq 0$,

3) $\displaystyle\int_0^{+\infty} u(t)\, dt < +\infty$.

Then $\lim\limits_{t\to\infty} u(t) = 0$.

Proof. Consider the following identity

$$u(t)^2 = u(0)^2 + 2\int_0^t u(\tau)\dot u(\tau)\, d\tau \tag{1.50}$$

and obtain the estimation

$$\int_0^t |u(\tau)\dot u(\tau)|\, d\tau \leq \int_0^t |\dot u(\tau)|\, u(\tau)\, d\tau \leq C\int_0^{+\infty} u(\tau)\, d\tau.$$

This estimate and assumption 3) of the lemma results in a convergence of the following integral

$$\int_0^{+\infty} u(\tau)\dot{u}(\tau)\,d\tau.$$

Then from (1.50) it follows that there exists a limit

$$\lim_{t\to+\infty} u(t)^2 = \nu.$$

Assumptions 2) and 3) imply that $\nu = 0$. The lemma is proved.

Lemma 1.4. *Let for a continuous on $[0, +\infty)$ function $u(t)$ the following assumptions hold:*

1) *for a number C we have*

$$\left|\ddot{u}(t)\right| \leq C, \qquad \forall\, t \geq 0,$$

2) $\lim_{t\to+\infty} u(t) = 0.$

Then $\lim_{t\to+\infty} \dot{u}(t) = 0.$

Proof. Suppose the opposite, i.e., there exists a sequence $t_k \to +\infty$ such that

$$|\dot{u}(t_k)| \geq \varepsilon.$$

In this case by 1) we obtain

$$|\dot{u}(t)| \geq \frac{\varepsilon}{2} \tag{1.51}$$

on the segments $\left[t_k,\ t_k + \frac{\varepsilon}{2C}\right]$.

It follows from assumption 2) that we may take t_1 such that

$$|u(t)| \leq \frac{\varepsilon^2}{16C}, \qquad \forall\, t \geq t_1. \tag{1.52}$$

By (1.51)

$$\left|u\left(t_k + \frac{\varepsilon}{2C}\right)\right| \geq \frac{3\varepsilon^2}{16C}.$$

The last inequality contradicts inequality (1.52). The proof of Lemma 1.4 is completed.

Theorem 1.4 (on transient process). *Suppose, the parameters of governor satisfy the following assumptions*

$$m\gamma_0 > \alpha^2, \tag{1.53}$$

$$\alpha\gamma_0 J > -\sqrt{3}\ F_0 f_0 m. \tag{1.54}$$

Then for a solution of equation (1.43) *with initial data* $\omega(0) = 0,\ \ \Delta x(0) = -x_0,\ \ (\Delta x(0))^\bullet = 0$ *the following relations*

$$\omega(t) \in \left[0,\ \sqrt{\frac{3\gamma_0 x_0}{\beta m r}}\ \right] = \left[0,\ \sqrt{3}\,\omega_0\right], \tag{1.55}$$

$$\lim_{t\to+\infty} \omega(t) = \omega_0, \qquad \lim_{t\to+\infty} \Delta x(t) = 0, \qquad \lim_{t\to+\infty} (\Delta x(t))^\bullet = 0 \tag{1.56}$$

are satisfied.

Recall that here $f_0 = 2\beta m\omega_0(x_0 + r)$.

Proof. Inclusion (1.55) follows directly from Lemma 1.1.

Really, ω_1 can be computed, using the equation

$$\frac{1}{3}\beta m r\omega_1^3 - \gamma_0 x_0\omega_1 = 0.$$

Condition (1.47) takes the form

$$\frac{\alpha\gamma_0}{m^2} > -\frac{2F_0}{mJ}\beta m r\omega_1$$

and is given by

$$\alpha\gamma_0 J > -\sqrt{3}\ F_0 m(f_0 - 2\beta m\omega_0 x_0).$$

The last inequality follows directly from condition (1.54).

Thus all the assumptions of Lemma 1.1 are satisfied and, consequently, inclusion (1.55) is valid.

To prove (1.56), note that by using (1.45) and the inclusion (1.55) we obtain

$$\dot V(\omega(t), y(t), z(t)) \le -\varepsilon y(t)^2, \qquad \forall\, t \ge 0, \tag{1.57}$$

where ε is a sufficiently small positive number. From (1.55) it follows that the function $V(\omega(t), y(t), z(t))$ is uniformly bounded on $[0, +\infty)$. Then by

(1.57) we obtain the existence of a number C such that

$$\int_0^t y(\tau)^2 d\tau \leq \frac{1}{\varepsilon}\left(V(\omega(0), y(0), z(0)) - V(\omega(t), y(t), z(t))\right) \leq C, \quad \forall t \geq 0.$$

The uniform boundedness of $V(\omega(t), y(t), z(t))$ and inclusion (1.55) result in the fact that on $[0, +\infty)$ the functions $y(t)$ and $z(t)$ are uniformly bounded. This implies that the function $\frac{d}{dt}y(t)^2 = 2y(t)z(t)$ is also uniformly bounded.

Thus, all the assumptions of Lemma 1.3 are satisfied and, consequently, we have

$$\lim_{t\to+\infty} y(t) = 0. \tag{1.58}$$

Let us also remark that

$$\ddot{y}(t) = -az(t) - by(t) - \varphi(\omega(t))$$

and, as it was proved above, $z(t), y(t), \omega(t)$ are uniformly bounded on $[0, +\infty)$. From (1.58) by Lemma 1.4 we conclude that for $z(t) = \dot{y}(t)$ the following relation

$$\lim_{t\to+\infty} z(t) = 0 \tag{1.59}$$

is satisfied. By (1.57)-(1.59) and using the special form of function V we obtain the existence of the following limits

$$\lim_{t\to+\infty} V(\omega(t), y(t), z(t)), \qquad \lim_{t\to+\infty} \int\limits_0^{\omega(t)} \varphi(x)\, dx.$$

Hence

$$\lim_{t\to+\infty} \omega(t) = \omega_0.$$

This completes the proof of the theorem.

Now we compare non-local conditions (1.53) and (1.54) of the transition from the initial state $\omega = 0, x = 0, \dot{x} = 0$ to the operating regime $\omega = \omega_0, x = x_0, \dot{x} = 0$ with conditions (1.41) of asymptotic stability of the operating regime.

Conditions (1.41) and (1.54) are similar in form. However, there is a slight difference: the expression on the right-hand side of inequality (1.54) includes a factor $\sqrt{3}$.

Condition (1.53) is an additional condition on the spring constant. Unlike conditions (1.41), the violation of which implies the physical nonrealizability of an operating regime (compare with the instability condition (1.42)), conditions (1.53) and (1.54) are sufficient conditions only. However in the engineering practice it is often impossible to pinpoint all parameters and the mathematical model is always a certain idealization. Therefore, in many cases the information, obtained by means of the sufficient conditions (1.53), (1.54), turns out to be quite sufficient.

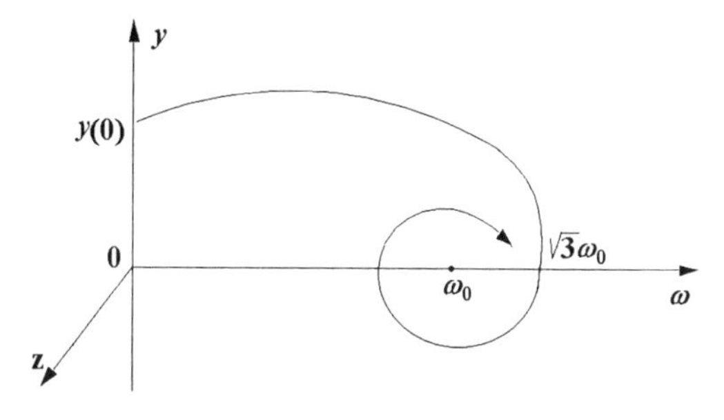

Fig. 1.11

The schematic representation of transient process is given in Fig. 1.11.

Chapter 2

Linear electric circuits. Transfer functions and frequency responses of linear blocks

2.1 Description of linear blocks

In the previous chapter a nonlinear mathematical model was considered. The linearization was an substantial part of investigation. In the present chapter we show that the widespread electric circuits, containing resistors, capacitors, and inductors, can be described by linear differential equations.

We consider first the simplest electrical circuit, RC-circuit, which is often used in radiotechnology as a low-pass filter (Fig. 2.1). Here R is a resistance, C is a capacitance, $u_1(t)$ and $u_2(t)$ are voltages.

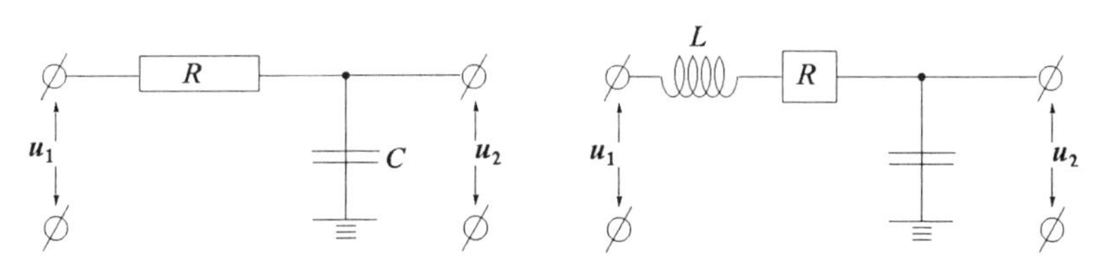

Fig. 2.1 Fig. 2.2

Let us find the relation between $u_1(t)$ and $u_2(t)$, using of the Ohm law

$$R\,i(t) = u_1(t) - u_2(t). \tag{2.1}$$

Here $i(t)$ is a current intensity. Recall that

$$i(t) = \frac{dq(t)}{dt}, \tag{2.2}$$

where $q(t)$ is a quantity of electricity. We can consider this quantity of electricity on the capacitor plates with the capacity C. Since the plate-to-plate voltage is $u_2(t)$, from the property of capacity we have

$$q(t) = Cu_2(t). \tag{2.3}$$

Putting (2.3) in (2.2) and (2.2) in (2.1), we obtain

$$RC\frac{du_2}{dt} + u_2 = u_1. \tag{2.4}$$

Consider now RLC–circuit (Fig. 2.2).

In this case an electromotive force of self-induction $e(t)$ is added to the voltage drop $u_1(t) - u_2(t)$. For $e(t)$ the following formula

$$e(t) = -L\frac{d\,i(t)}{dt} \tag{2.5}$$

is well known. Therefore in place of relation (2.1) we can take

$$u_1(t) - u_2(t) + e(t) = R\,i(t) \tag{2.6}$$

or, using (2.5), the following equation

$$u_1(t) - u_2(t) - L\frac{d\,i(t)}{dt} = R\,i(t). \tag{2.7}$$

Formulas (2.2) and (2.3) are obviously valid. Substituting (2.3) into (2.2) and (2.2) in (2.7), we finally obtain

$$LC\frac{d^2u_2}{dt^2} + RC\frac{du_2}{dt} + u_2 = u_1. \tag{2.8}$$

Thus, equations (2.4) and (2.8) are those relating u_1 and u_2 for the RC and RCL-circuits respectively. It is also useful to regard the values $u_1(t)$ and $u_2(t)$ as the input and output of the block respectively. These blocks are described by equations (2.4) or (2.8) (Fig. 2.3).

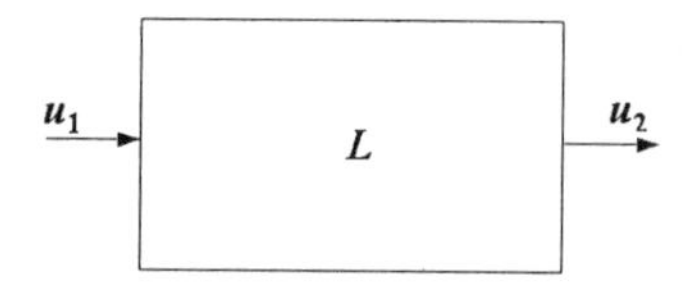

Fig. 2.3

Let us remark that from the formal point of view in both cases the input and output can be interchanged: it is possible to supply voltage $u_1(t)$ on the input and then to observe voltage $u_2(t)$ on the output and vice versa. However $u_2(t)$ as the output of block L turns out to be a solution of equations (2.4) with initial data $u_2(0)$ or with initial data $u_2(0)$, $\dot{u}_2(0)$ while the output $u_1(t)$ is uniquely described by formulas (2.4) and (2.8) only.

Note also that in an engineering practice the cases that a linear block is a sum of operators of differentiation are rare in occurrence. In this case to an input $u_2(t) + A \sin \omega t$ (A is small and ω is large) assign the output

$$u_1(t) = RC\dot{u}_2(t) + u_2(t) + RCA\omega \cos \omega t + A \sin \omega t.$$

Since in this case the value $RCA\omega$ is not small, the signal $u_2(t)$ passes through the block L with a large distortion. Further we shall show that in the case that an input is $u_1(t)$ there occurs an inverse effect and high-frequency distortions of the kind $A \sin \omega t$ are depressed.

Since formulas (2.1)–(2.3) and (2.5), (2.6) are also valid for the other of electric circuits, containing conductors, resistors, capacitors, and inductors, these circuits are described by linear differential equations with constant coefficients. In this case we are interested in the answer to the question of how the signal $u_1(t)$ is changed when passing via linear circuit L, i.e., how the input $u_1(t)$ and the output $u_2(t)$ of linear block L are related to one another.

To answer this question in the framework of the linear circuits theory, at the end of the 19th century and early in the 20s the fundamental concepts of the control theory such as input, output, transfer function, and frequency response were stated.

We see that equations (2.4) and (2.8) admit a natural generalization:

$$\mathcal{N}\left(\frac{d}{dt}\right) u_2 = \mathcal{M}\left(\frac{d}{dt}\right) u_1. \tag{2.9}$$

Here $\mathcal{N}(\frac{d}{dt})$ and $\mathcal{M}(\frac{d}{dt})$ are the following differential operators

$$\mathcal{N}\left(\frac{d}{dt}\right) u := \mathcal{N}_n u^{(n)} + \mathcal{N}_{n-1} u^{(n-1)} + \ldots + \mathcal{N}_1 \dot{u} + \mathcal{N}_0 u,$$
$$\mathcal{M}\left(\frac{d}{dt}\right) u := \mathcal{M}_m u^{(m)} + \mathcal{M}_{m-1} u^{(m-1)} + \ldots + \mathcal{M}_1 \dot{u} + \mathcal{M}_0 u,$$

where $\mathcal{N}_i$ and $\mathcal{M}_i$ are some numbers. The previous remark on the noise immunity of blocks (2.4) and (2.8) implies the following restriction: $m < n$. Without loss of generality it can be assumed that $\mathcal{N}_n = 1$.

The following notation of inputs and outputs $\sigma = u_2$, $\xi = u_1$ are needed for the sequel.

We discuss once more the transformation of a signal $\xi(t)$ by the linear block L (Fig. 2.4) if this block is given by means of the following equation

$$\mathcal{N}\left(\frac{d}{dt}\right)\sigma = \mathcal{M}\left(\frac{d}{dt}\right)\xi. \tag{2.10}$$

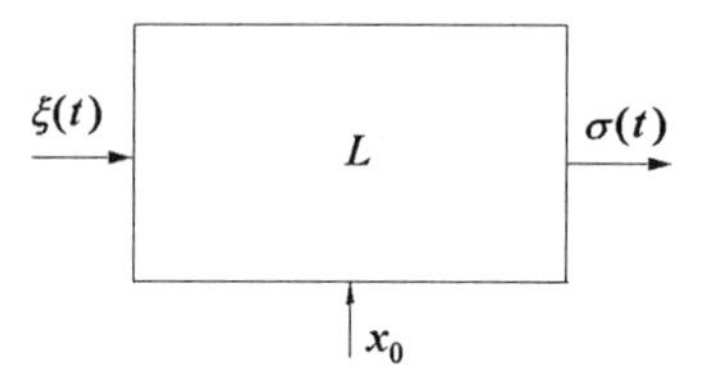

Fig. 2.4

The operator $M\left(\frac{d}{dt}\right)$ acts first on the function $\xi(t)$:

$$f(t) = \mathcal{M}\left(\frac{d}{dt}\right)\xi(t).$$

Then $\sigma(t)$ is defined as a solution of a nonhomogeneous linear equation

$$\mathcal{N}\left(\frac{d}{dt}\right)\sigma(t) = f(t). \tag{2.11}$$

It is clear that $\sigma(t)$ is not determined by the function $\xi(t)$ only. The initial data are the following

$$x_0 = \begin{pmatrix} \sigma(0) \\ \dot{\sigma}(0) \\ \vdots \\ \sigma^{(n-1)}(0) \end{pmatrix}.$$

The vector x_0 is called initial data of block L.

So, the block L is the operator, described above, which acts on the direct product of sets

$$\{\xi(t)\} \times \{x_0\} \xrightarrow{L} \{\sigma(t)\}. \tag{2.12}$$

The set of inputs $\{\xi(t)\}$ is a set of functions, which the operator L can be defined on. In the case considered it is the functions that at any point $t \in [0, +\infty)$ have m continuous derivatives. Then the function

$f(t) = \mathcal{M}(\frac{d}{dt})\xi(t)$ is defined and continuous at any point $t \in [0, +\infty)$. The continuity of $f(t)$ implies the existence of the solution $\sigma(t)$ of equation (2.11), which is determined for all $t \geq 0$ and has the n–th continuous derivative. The set of initial data is a subset of Euclidean space.

Now we discuss another description of block L, which makes it possible to define the operator L on a set of continuous functions $\{\xi(t)\}$ only. Using previous notation, the operator L is as follows

$$\mathcal{N}\left(\frac{d}{dt}\right)\eta = \xi, \qquad \sigma = \mathcal{M}\left(\frac{d}{dt}\right)\eta, \qquad x_0 = \begin{pmatrix} \eta(0) \\ \dot{\eta}(0) \\ \vdots \\ \eta^{(n-1)}(0) \end{pmatrix}. \tag{2.13}$$

We define a function $\eta(t)$ as a solution of the equation

$$\mathcal{N}\left(\frac{d}{dt}\right)\eta = \xi$$

with initial data x_0 and apply $M(\frac{d}{dt})$ to the function $\eta(t)$: $\sigma(t) = M(\frac{d}{dt})\eta(t)$.

It can readily be seen that if the block L is given by equations (2.13) and a function $\xi(t)$ has the m–th continuous derivative, then it is possible to describe this block by equations (2.10). Really, we have

$$\mathcal{N}\left(\frac{d}{dt}\right)\sigma = \mathcal{N}\left(\frac{d}{dt}\right)\mathcal{M}\left(\frac{d}{dt}\right)\eta =$$

$$= \mathcal{M}\left(\frac{d}{dt}\right)\mathcal{N}\left(\frac{d}{dt}\right)\eta = \mathcal{M}\left(\frac{d}{dt}\right)\xi.$$

Now we shall show that description (2.13) is a partial case of the following equations

$$\begin{aligned} \frac{dx}{dt} &= Ax + b\xi, \\ & \qquad\qquad x_0 = x(0), \\ \sigma &= c^* x, \end{aligned} \tag{2.14}$$

where A is a constant $n \times n$-matrix, b and c are constant matrices of dimensions $n \times m$ and $n \times l$ respectively. The asterisk $*$ denotes the transposition in the real case and Hermitian conjugation in the complex case.

For the sequel we are needed in the following notation

$$\eta = x_1, \quad \dot{\eta} = x_2, \quad \dots, \eta^{(n-1)} = x_n, \quad x = \begin{pmatrix} x_1 \\ \vdots \\ x_n \end{pmatrix}.$$

Then equations (2.13) may be written by

$$\dot{x}_1 = x_2,$$

$$\dots\dots\dots\dots\dots\dots\dots\dots\dots\dots\dots\dots\dots$$

$$\dot{x}_{n-1} = x_n,$$
$$\dot{x}_n = -\mathcal{N}_{n-1}x_n - \dots - \mathcal{N}_0 x_1 + \xi(t),$$

$$\sigma = \mathcal{M}_m x_{m+1} + \mathcal{M}_{m-1}x_m + \dots + \mathcal{M}_0 x_1.$$

Hence we have

$$A = \begin{pmatrix} 0 & 1 & 0 & \dots & 0 \\ 0 & 0 & 1 & \dots & \ddots \\ \vdots & \vdots & \ddots & \ddots & 0 \\ 0 & 0 & \dots & 0 & 1 \\ -\mathcal{N}_0 & \dots & \dots & \dots & -\mathcal{N}_{n-1} \end{pmatrix}, \qquad b = \begin{pmatrix} 0 \\ \vdots \\ \vdots \\ 0 \\ 1 \end{pmatrix},$$

$$c = \begin{pmatrix} \mathcal{M}_0 \\ \vdots \\ \mathcal{M}_m \\ 0 \\ 0 \end{pmatrix}, \quad x_0 = \begin{pmatrix} x_1(0) \\ \vdots \\ x_n(0) \end{pmatrix} = \begin{pmatrix} \eta(0) \\ \vdots \\ \eta^{(n-1)}(0) \end{pmatrix}.$$

Let us recall that a solution of equation (2.14) may be written in the following integral form (the Cauchy formula)

$$x(t) = e^{At}x_0 + \int_0^t e^{A(t-\tau)}b\xi(\tau)d\tau.$$

Therefore

$$\sigma(t) = c^* e^{At}x_0 + \int_0^t c^* e^{A(t-\tau)}b\xi(\tau)\,d\tau. \tag{2.15}$$

Given the description of block L in the form (2.15), the following generalization can be obtained

$$\sigma(t) = \alpha(t) + \int_0^t \gamma(t, \tau)\, \xi(\tau)\, d\tau. \tag{2.16}$$

Here $\alpha(t)$ is a continuous l-dimension vector-function from a certain functional set $\{\alpha(t)\}$. In the theory of integral operators a matrix function $\gamma(t, \tau)$ of dimension $l \times m$ is called a kernel of integral operator

$$\int_0^t \gamma(t, \tau)\, \xi(\tau)\, d\tau.$$

The description of blocks (2.14) can also be transformed to the form (2.16) if matrices A and b depend on t.

Further we consider the case of a difference kernel only

$$\gamma(t, \tau) = \gamma(t - \tau).$$

The following transformation

$$\int_0^t \gamma(t - \tau)\xi(\tau)\, d\tau$$

is often called a convolution operator.

Thus, among the descriptions of L, operator (2.16) is most general. In the case of description (2.14) we obtain

$$\alpha(t) = c^* e^{At} x_0, \qquad \gamma(t, \tau) = c^* e^{A(t-\tau)} b.$$

Now we discuss the answer to the question: In what sense is there understood a linearity of all the above-mentioned descriptions of block L? In the case of descriptions (2.10), (2.13), and (2.14) the block is linear since all the equations mentioned are linear. However it is more convenient to introduce another definition of linearity that includes description (2.16).

Definition 2.1. *A block L is called linear if to any linear combination of arbitrary inputs $\xi_1(t)$ and $\xi_2(t)$*

$$\mu_1\xi_1(t) + \mu_2\xi_2(t)$$

assigns the linear combination

$$\mu_1(\sigma_1(t) - \alpha_1(t)) + \mu_2(\sigma_2(t) - \alpha_2(t)).$$

Here

$$\sigma_i(t) = \alpha_i(t) + \int_0^t \gamma(t,\tau)\,\xi_i(\tau)\,d\tau.$$

Since descriptions (2.10), (2.13), and (2.14) of the block L may be reduced to the form (2.16) and description (2.16) is obviously linear, the expressions (2.10), (2.13), and (2.14) are also linear.

2.2 Transfer functions and frequency responses of linear blocks

For the linear blocks, described by equations (2.10) or (2.13) we define a transfer function in the following way.

Definition 2.2. *A function*

$$W(p) = -\frac{\mathcal{M}(p)}{\mathcal{N}(p)}, \tag{2.17}$$

defined on a complex plane $\mathbb{C}$, *is called a transfer function* $W(p)$ *of linear block.*

Definition 2.3. *An* $l \times m$*-matrix*

$$W(p) = c^*(A - pI)^{-1}b, \tag{2.18}$$

whose elements are the functions defined on a complex plane $\mathbb{C}$, *is called a transfer function (matrix function) of a linear block given by equations* (2.14).

It is clear that transfer function (2.17) is defined everywhere except for the points $\mathbb{C}$ that are zeros of polynomial $\mathcal{N}(p)$. These singular points are poles of function $W(p)$. The same proposition is valid for $W(p)$ in the form (2.18). By using the rule of matrix inversion we conclude that elements of matrix $(A - pI)^{-1}$ take the form

$$\frac{\alpha_{ij}(p)}{\alpha(p)}, \quad i = 1, \ldots, n, \; j =, \ldots, n,$$

where $\alpha(p)$ is a characteristic polynomial of matrix A: $\alpha(p) = \det(pI - A)$ and $\alpha_{ij}(p)$ are some polynomials of degree not greater than $n - 1$. In this case the elements of $W(p)$ are the following functions

$$\frac{\beta_{ij}(p)}{\alpha(p)}, \quad i = 1, \ldots, l, \; j =, \ldots, m$$

where $\beta_{ij}(p)$ are some polynomials of degree not greater than $n - 1$. The poles of these functions are zeros of polynomial $\alpha(p)$. In other words, the poles of $W(p)$ coincide with the eigenvalues of matrix A.

Show that functions (2.17) and (2.18) are well defined. It means that if we rewrite equations (2.10) and (2.13) in the form (2.14)., then the transfer functions from Definitions 2.2 and 2.3 coincide. In this case the matrices A, b, and c have the form

$$A = \begin{pmatrix} 0 & 1 & 0 & \ldots & 0 \\ 0 & 0 & 1 & \ddots & \vdots \\ \vdots & \vdots & \ddots & \ddots & 0 \\ 0 & 0 & \ldots & 0 & 1 \\ -\mathcal{N}_0 & \ldots & \ldots & \ldots & -\mathcal{N}_{n-1} \end{pmatrix}$$

$$b = \begin{pmatrix} 0 \\ \vdots \\ \vdots \\ 0 \\ 1 \end{pmatrix}, \qquad c = \begin{pmatrix} \mathcal{M}_0 \\ \vdots \\ \mathcal{M}_m \\ 0 \\ \vdots \\ 0 \end{pmatrix}. \tag{2.19}$$

For the matrices A, b, and c the following formula

$$c^*(A - pI)^{-1}b = -\frac{\mathcal{M}_0 + \mathcal{M}_1 p + \ldots + \mathcal{M}_m p^m}{\det(pI - A)} \tag{2.20}$$

is valid.

Indeed the type of vector b implies that for the expression $(A - pI)^{-1}b$ to be computed it is necessary to find the last column of the matrix $(A - pI)^{-1}$

only. Using the rule of matrix inversion we obtain that this column consists of algebraical complements of the last row of matrix, namely

$$-\begin{pmatrix} p & -1 & 0 & \dots & \dots & & 0 \\ 0 & p & -1 & 0 & \dots & & \vdots \\ \dots & \dots & \dots & \dots & \dots & \dots & 0 \\ 0 & \dots & \dots & \dots & \dots & p & -1 \\ \mathcal{N}_0 & \dots & \dots & \dots & \dots & \mathcal{N}_{n-2} & (\mathcal{N}_{n-1}+p) \end{pmatrix},$$

divided by $\det(pI - A)$.

It is obvious that the required algebraical complements are equal to $-1, -p, \dots, -p^{n-1}$. Therefore

$$(A - pI)^{-1}b = -\frac{1}{\det(pI - A)} \begin{pmatrix} 1 \\ p \\ \vdots \\ p^{n-1} \end{pmatrix}.$$

Formula (2.20) follows directly from the last relation.

Since

$$\det(pI - A) = p^n + \mathcal{N}_{n-1}p^{n-1} + \dots + \mathcal{N}_0,$$

we obtain for the matrices of the form (2.19) the following formula

$$c^*(A - pI)^{-1}b = -\frac{\mathcal{M}(p)}{\mathcal{N}(p)}. \tag{2.21}$$

Note a very important property of a transfer matrix of system (2.14).

Theorem 2.1. *The transfer matrix $W(p)$ is invariant with respect to nonsingular linear transformation $x = Sy$ $(\det S \neq 0)$.*

Proof. Having performed the linear nonsingular transformation $x = Sy$ in system (2.14), we obtain a new system

$$\frac{dy}{dt} = S^{-1}ASy + S^{-1}b\xi,$$
$$\sigma = c^*Sy. \tag{2.22}$$

Let us compute a transfer matrix of this system

$$W_1(p) = c^*S(S^{-1}AS - pI)^{-1}S^{-1}b.$$

Since

$$(S^{-1}AS - pI)^{-1} = (S^{-1}AS - pS^{-1}S)^{-1} =$$
$$= (S^{-1}(A - pI)S)^{-1} = S^{-1}(A - pI)^{-1}S,$$

we have

$$c^* S(S^{-1}AS - pI)^{-1}S^{-1}b = c^*(A - pI)^{-1}b.$$

Thus, the transfer matrices $W(p)$ of system (2.14) and $W_1(p)$ of system (2.22) coincide.

The correctness of Definitions 2.2 and 2.3 follows directly from formula (2.21) and the invariance of transfer functions of system (2.14).

Consider now a linear block of the form (2.16) with difference kernel

$$\gamma(t, \tau) = \gamma(t - \tau).$$

To define the transfer function of such a block we need the Laplace transformation. Consider a set of real continuous on $[0, +\infty)$ functions $\{f(t)\}$ satisfying the following inequalities

$$|f(t)| \leq \rho e^{æt}, \qquad \forall\, t \geq 0. \tag{2.23}$$

Here the numbers ρ can be different for different functions from the set $\{f(t)\}$ and the numbers æ are the same for the whole set $\{f(t)\}$.

Definition 2.4. *The Laplace transformation is an operator, defined on the set $\{f(t)\}$, such that it takes each element from this set to an element of the set of complex-valued functions $\{g(p)\}$, given on the set*

$$\{p \mid \ p \in \mathbb{C}, \quad \mathrm{Re}\, p > æ\}, \tag{2.24}$$

by the following rule

$$g(p) = \int_0^{+\infty} e^{-pt} f(t)\, dt. \tag{2.25}$$

Inequality (2.23) implies, obviously, the convergence of integral (2.25).

This definition may be extended in the natural way to the vector functions $f(t)$. In this case on the left of inequality (2.23) Euclidean norm $|\cdot|$ must be taken rather than the absolute value $|\cdot|$ and $g(p)$ are vector functions, whose dimensions are the same as that of vector functions $f(t)$.

Let us now state two important properties of the Laplace operator, denoting this operator by L: $\{f(t)\} \to \{g(p)\}$.

Proposition 2.1. *If a function $f(t)$ from a set $\{f(t)\}$ has a continuous derivative at any point t, then for $\dot{f}(t)$ the Laplace operator may be defined by the formula*

$$L(\dot{f}(t)) = \int_0^{+\infty} e^{-pt} \dot{f}(t)\, dt$$

and the following relation

$$L(\dot{f}(t)) = pL(f(t)) - f(0) \tag{2.26}$$

is valid.

The proof of this proposition is of a chain of relations

$$\int_0^{+\infty} e^{-pt} \dot{f}(t)\, dt = e^{-pt} f(t)\Big|_0^{+\infty} + p\int_0^{+\infty} e^{-pt} f(t)\, dt = -f(0) + pL(f(t)).$$

Thus, the Laplace operator may be extended from the set $\{f(t)\}$ to the set $\{f(t)\} \cup \{\dot{f}(t)\}$.

We recall also that the integral

$$\int_0^{+\infty} e^{-pt} \dot{f}(t)\, dt$$

does not always converge absolutely.

Proposition 2.2. *For functions $f_1(t)$ and $f_2(t)$ from the set $\{f(t)\}$ the following formula*

$$L\left(\int_0^t f_1(t-\tau)\, f_2(\tau)\, d\tau\right) = L(f_1(t))\, L(f_2(t)) \tag{2.27}$$

is valid.

The proof is of the following chain of relations

$$L\Big(\int_0^t f_1(t-\tau)f_2(\tau)\,d\tau\Big) = \int_0^{+\infty} e^{-pt}\int_0^t f_1(t-\tau)f_2(\tau)\,d\tau\,dt =$$

$$= \iint_\Omega e^{-pt} f_1(t-\tau)\,f_2(\tau)\,dt\,d\tau = \int_0^{+\infty} f_2(\tau)\int_\tau^{+\infty} e^{-pt} f_1(t-\tau)\,dt\,d\tau =$$

$$= \int_0^{+\infty} e^{-p\tau} f_2(\tau)\int_0^{+\infty} e^{-pt_1} f_1(t_1)\,dt_1\,d\tau =$$

$$= \int_0^{+\infty} e^{-p\tau} f_2(\tau)\,d\tau\int_0^{+\infty} e^{-pt_1} f_1(t_1)\,dt_1 = L(f_2(t))\,L(f_1(t)).$$

Here $t_1 = t - \tau$ and the domain Ω has the form shown in Fig. 2.5.

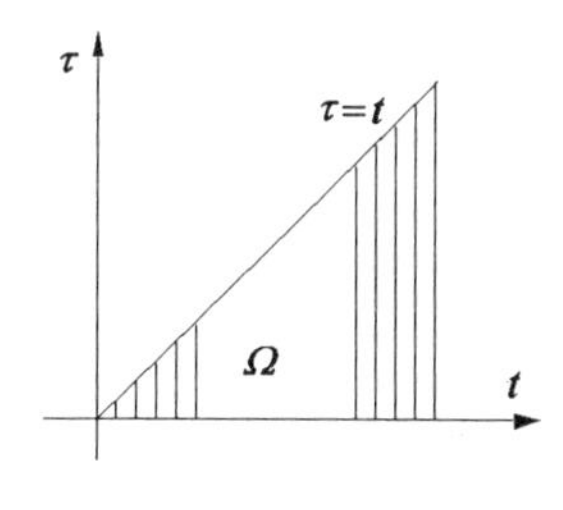

Fig. 2.5

All the integrals in these relations converge absolutely by virtue of inequality (2.23).

Propositions 2.1 and 2.2 are also true for the set of matrix functions or vector functions $\{f(t)\}$. In this case in formula (2.27) $f_1(t)$ is a $k\times m$-matrix function and $f_2(t)$ is an $m\times l$-matrix function.

Let us apply the tools, developed above, to the analysis of equations (2.13) and (2.14). Note that if an input $\xi(t)$ satisfies the inequality

$$|\xi(t)| \le \rho_1 e^{æ_1 t}, \qquad \forall\, t \ge 0,$$

then for the vector function $x(t)$ and the function $\eta(t)$, numbers $æ_2$ and ρ_2

can be found such that inequalities

$$|x(t)| \le \rho_2 e^{æ_2 t}, \quad |\eta^{(i)}(t)| \le \rho_2 e^{æ_2 t}, \qquad \forall\, t \ge 0 \tag{2.28}$$

are satisfied. Here $i = 0, 1, \ldots, n-1$.

For proving this fact it is sufficient to show that the first inequality holds. The latter follows directly from estimates

$$|x(t)| \le |e^{At}x(0)| + \left| \int_0^t e^{A(t-\tau)} b\xi(\tau)\, d\tau \right| \le$$

$$\le \beta e^{\lambda t}|x(0)| + \int_0^t |e^{A(t-\tau)}||b||\xi(\tau)|\, d\tau \le$$

$$\le \beta e^{\lambda t}|x(0)| + \beta |b|\, \rho_1 e^{\lambda t} \int_0^t e^{(æ_1 - \lambda)\tau} d\tau.$$

Here numbers β and λ are such that

$$|e^{At}| \le \beta e^{\lambda t}, \quad \forall\, t \ge 0.$$

It is obvious that estimates (2.28) imply the existence of ρ_3 such that

$$|\sigma(t)| \le \rho_3 e^{æ_2 t}, \qquad \forall\, t \ge 0. \tag{2.29}$$

Estimates (2.28) and (2.29) can be obtained by different methods, they can also be improved, and so on. In our consideration an important point is that estimates of this kind exist.

In this case, choosing $æ = \max(æ_1, æ_2)$ and assuming zero initial data: $x(0) = 0$ or $\eta^{(i)}(0) = 0, \;\; i = 0, 1, \ldots, n-1$, by Proposition 2.1 we obtain the following relations

1) for equations (2.13)

$$\left.\begin{aligned} L\left(\mathcal{N}\left(\frac{d}{dt}\right)\eta(t)\right) &= L(\xi(t)) \\ L(\sigma(t)) &= L\left(\mathcal{M}\left(\frac{d}{dt}\eta(t)\right)\right) \end{aligned}\right\}$$

$$\left.\begin{aligned} \mathcal{N}(p)\, L(\eta(t)) &= L(\xi(t)) \\ L(\sigma(t)) &= \mathcal{M}(p)\, L(\eta(t)) \end{aligned}\right\} \tag{2.30}$$

$$L(\sigma(t)) = \frac{\mathcal{M}(p)}{\mathcal{N}(p)}\, L(\xi(t)),$$

$$L(\sigma(t)) = -W(p)\, L(\xi(t))$$

2) for equations (2.14)

$$\left.\begin{aligned} L(\dot{x}(t)) &= L(Ax(t) + b\xi(t)) \\ L(\sigma(t)) &= L(c^* x(t)) \end{aligned}\right\}$$

$$\left.\begin{aligned} pL(x(t)) &= AL(x(t)) + bL(\xi(t)) \\ L(\sigma(t)) &= c^*\, L(x(t)) \end{aligned}\right\}$$

$$\left.\begin{aligned} L(x(t)) &= -(A - pI)^{-1} bL(\xi(t)) \\ L(\sigma(t)) &= c^*\, L(x(t)) \end{aligned}\right\} \tag{2.31}$$

$$L(\sigma(t)) = -c^*(A - pI)^{-1} bL(\xi(t)),$$

$$L(\sigma(t)) = -W(p)\, L(\xi(t)).$$

Thus, we obtain a very simple relation between the Laplace transforms of the input and output of linear block.

If we now consider a description of linear block (2.16) with the difference kernel $\gamma(t, \tau) = \gamma(t - \tau)$:

$$\sigma(t) = \alpha(t) + \int_0^t \gamma(t - \tau)\, \xi(\tau)\, d\tau, \tag{2.32}$$

assuming that the initial data of this block are such that $\alpha(t) \equiv 0$, then by Proposition 2.2 we obtain a relation similar to relations

(2.30) and (2.31)

$$L(\sigma(t)) = L(\gamma(t))\, L(\xi(t)). \tag{2.33}$$

In this case the following definition can be stated.

Definition 2.5. *The Laplace transform of $\gamma(t)$, taken with the opposite sign,*

$$W(p) = -L(\gamma(t)) \tag{2.34}$$

is called a transfer function (matrix function) $W(p)$ of the linear block, described by equations (2.32).

For the function $\gamma(t) = c^* e^{At} b$ we have

$$L(\gamma(t)) = L(c^* e^{At} b) = \int_0^{+\infty} e^{-pt} c^* e^{At} b\, dt =$$

$$= c^* (A - pI)^{-1} e^{(A-pI)t} b \Big|_0^{+\infty} = -c^* (A - pI)^{-1} b.$$

Here we take into account the following relation

$$\lim_{t \to +\infty} e^{(A-pI)t} = 0,$$

which is valid under the assumption that $\operatorname{Re} p > æ > \max_j \operatorname{Re} \lambda_j(A)$, were $\lambda_j(A)$ are eigenvalues of matrix A. Therefore function (2.34) is well defined.

Let us now proceed to the definition of frequency response of linear block. Suppose that $æ < 0$ and, consequently, the transfer function (matrix function) is also defined on the imaginary axis.

Definition 2.6. *A function $W(i\omega)$ is called a frequency response of linear block.*

The Fourier transformation, together with the Laplace transformation, plays an important role in many fields of applied mathematics.

In the control theory the Fourier transformation is often defined in such a way that it coincides with the Laplace transformation on the imaginary axis

$$F(f(t)) = \int_0^{+\infty} e^{-i\omega t} f(t)\, dt. \tag{2.35}$$

Such a definition, for example, can be found in [26].

In the calculus (see [14]) the Fourier transformation of the functions $f(t)$, given on $(-\infty, +\infty)$, is defined in the following way

$$\mathcal{F}(f(t)) = \frac{1}{\sqrt{2\pi}} \int_{-\infty}^{+\infty} e^{i\omega t} f(t)\, dt. \tag{2.36}$$

We assume that $f(t) \equiv 0$ on $(-\infty, 0)$. In this case the transforms $(F(f))(i\omega)$ and $(\mathcal{F}(f))(i\omega)$ of the same function $f(t)$ are related by the formula

$$(F(f))(i\omega) = \sqrt{2\pi}\ (\mathcal{F}(f))(-i\omega). \tag{2.37}$$

Such a difference in definitions is, as a rule, unessential and the use of one or the other of definition is a matter of convenience only. Further we shall use definition (2.35).

From formulas (2.30), (2.31), and (2.33) it follows that for zero initial data of block we have

$$F(\sigma(t)) = -W(i\omega)\, F(\xi(t)). \tag{2.38}$$

Definition 2.7. *A hodograph of frequency response $W(i\omega)$ (Nyquist plot) is a set of all its values on the complex plane $\mathbb{C}$.*

In Fig. 2.6 is shown an example of hodograph $W(i\omega)$. The arrows on the hodograph indicate the direction in which $W(i\omega)$ changes with increasing ω.

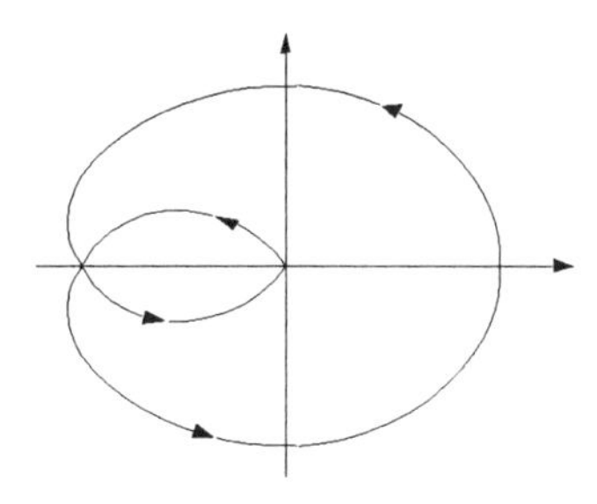

Fig. 2.6

Consider the case of a harmonic input $\xi(t)$. We restrict ourselves by descriptions of blocks (2.13) and (2.14) for $m = l = 1$. We shall seek solutions $\eta(t)$ and $x(t)$ in the form $\eta(t) = V(i\omega)e^{i\omega t}$, $x(t) = U(i\omega)e^{i\omega t}$.

Here $V(i\omega)$ is a scalar and $U(i\omega)$ is a vector. Substituting these expressions into equations (2.13) and (2.14), we obtain

$$V(i\omega)\,\mathcal{N}(i\omega)e^{i\omega t} = e^{i\omega t},$$
$$(A - i\omega I)\,U(i\omega)e^{i\omega t} + be^{i\omega t} = 0.$$

These equations are satisfied in the case that

$$V(i\omega) = \frac{1}{\mathcal{N}(i\omega)}, \qquad U(i\omega) = -(A - i\omega I)^{-1}b. \tag{2.39}$$

Thus, there exist harmonic solutions $\eta(t)$, $x(t)$ and the output $\sigma(t)$ takes the form

$$\sigma(t) = \frac{\mathcal{M}(i\omega)}{\mathcal{N}(i\omega)}\,e^{i\omega t},$$

$$\sigma(t) = -c^*(A - i\omega I)^{-1}be^{i\omega t}.$$

Introducing a frequency response $W(i\omega)$, finally we obtain

$$\sigma(t) = -W(i\omega)\,e^{i\omega t}. \tag{2.40}$$

Thus, for $\xi(t) = e^{i\omega t}$ there exist initial data of blocks such that the output $\sigma(t)$ is determined by formula (2.40).

Further we shall consider stable blocks only, namely block (2.13) provided that $\mathcal{N}(p)$ is a stable polynomial and block (2.14) provided that A is a stable matrix (i.e., all its eigenvalues have negative real parts). In this case for two solutions $x_1(t)$ and $x_2(t)$ with different initial data $x_1(0)$ and $x_2(0)$ we have

$$\big(x_1(t) - x_2(t)\big)^\bullet = A\big(x_1(t) - x_2(t)\big).$$

Taking into account the properties of the matrix A, we obtain

$$\lim_{t\to+\infty}\big(x_1(t) - x_2(t)\big) = 0.$$

Whence it follows that for the corresponding inputs $\sigma_1(t)$ and $\sigma_2(t)$ the following relation holds

$$\lim_{t\to+\infty}\big(\sigma_1(t) - \sigma_2(t)\big) = 0. \tag{2.41}$$

Since equations (2.13) are a special case of description (2.14), we see that (2.41) is valid for block description (2.13).

From (2.40) and (2.41) it follows that for stable blocks for $\xi(t) = e^{i\omega t}$ and arbitrary initial data we have

$$\lim_{t\to+\infty} \left(\sigma(t) + W(i\omega)e^{i\omega t}\right) = 0. \tag{2.42}$$

Formula (2.42) results in that for stable blocks the frequency response may be found experimentally. For this purpose it is necessary to apply a harmonic input $\xi(t) = e^{i\omega t}$ and then to wait some time until there occurs asymptotically harmonic output

$$\sigma(t) = A(i\omega)e^{i\omega t+\alpha(i\omega)i}.$$

Here $A(i\omega)$ is an amplitude of signal, $\alpha(i\omega)$ a phase shift. We have

$$A(i\omega) = |W(i\omega)|, \qquad \alpha(i\omega) = \pi + \arg W(i\omega).$$

These formulas determine uniquely a complex number $W(i\omega)$. By varying ω from $-\infty$ to $+\infty$, we obtain a hodograph $W(i\omega)$.

Thus, to find in this way the frequency response, we do not need information on the coefficients of polynomials $\mathcal{M}(p)$ and $\mathcal{N}(p)$ or that on the matrix A and vectors b and c. Since many results of the control theory are stated in terms of frequency responses, the descriptions of blocks by means of differential or integral equations are not necessary. In this case a curve on the complex plane, which is a hodograph of frequency response, is required only.

In some cases a frequency response makes it possible to determine uniquely the equations in question. Let us assume, for example, that, a priori, we know that polynomials $\mathcal{M}(p)$ and $\mathcal{N}(p)$ in description (2.13) do not have common zeros. Then, using $W(i\omega)$, one can uniquely determine $W(p)$ (by the principle of analytic continuation) and, using the function $W(p)$, the polynomials $\mathcal{M}(p)$ and $\mathcal{N}(p)$ can be found.

Chapter 3

Controllability, observability, stabilization

3.1 Controllability

In this chapter the study of linear systems will be continued.

Consider linear blocks, described by the following equations

$$\frac{dx}{dt} = Ax + b\xi,$$
$$\sigma = c^* x. \tag{3.1}$$

Here A is a constant $n \times n$-matrix, b is a constant $n \times m$-matrix, c is a constant $n \times l$-matrix, $\xi(t)$ is an m-dimensional vector function, which is regarded as an input of block. Thus, an output of the block $\sigma(t)$ is an l-dimensional vector function.

Definition 3.1. [23] *We shall say that system* (3.1) *is controllable if for any number $T > 0$ and any pair of vectors $x_0 \in \mathbb{R}^n$, $x_1 \in \mathbb{R}^n$ there exists a vector function $\xi(t)$ such that for a solution $x(t)$ of system* (3.1) *with both $\xi(t)$ and the initial data $x(0) = x_0$ the relation $x(T) = x_1$ is satisfied.*

Thus, system (3.1) is controllable if its state vector $x(t)$ can be transformed in a definite time T from an arbitrary initial data x_0 into a final state x_1 by the control $\xi(t)$.

Theorem 3.1. *The following conditions are equivalent and each of them is a necessary and sufficient condition of a controllability of system* (3.1) :

1) *a rank of an $n \times nm$-matrix $(b, Ab, \dots, A^{n-1}b)$ equals n:*

$$\mathrm{rank}(b, Ab, \dots, A^{n-1}b) = n, \tag{3.2}$$

1$'$) *a relation* $z^* A^k b = 0$, $\forall\, k = 0, \ldots, n-1$, *where* $z \in \mathbb{R}^n$, *implies that* $z = 0$,

2) *a linear space* L, *spanned by* n*-vectors being the columns of a matrix* $e^{At}b$ *for* $t \in \mathbb{R}^1$, *coincides with* $\mathbb{R}^n$*:*

$$L\{e^{At}b \,|\, t \in (-\infty, +\infty)\} = \mathbb{R}^n, \tag{3.3}$$

2$'$) *for arbitrary numbers* $\tau_1 < \tau_2$ *the following relation holds*

$$L\{e^{At}b \,|\, t \in (\tau_1, \tau_2)\} = \mathbb{R}^n, \tag{3.4}$$

3) *for arbitrary numbers* $\tau_1 < \tau_2$ *the following matrix is symmetric and positively definite:*

$$K = \int_{\tau_1}^{\tau_2} e^{-At}\, b\, b^* e^{-A^* t} dt > 0. \tag{3.5}$$

Before proving this theorem we recall that a linear space $L\{a_1, \ldots, a_k\}$, spanned by vectors $a_1, \ldots, a_k$, is a set of all possible linear combinations of these vectors $\{ \sum_{j=1}^{k} \alpha_j a_j | \alpha_j \in \mathbb{R}^1 \}$.

The columns $a_1, \ldots, a_k$ of the matrix $e^{At}b$ depend on t, namely $a_1(t), \ldots, a_k(t)$. Therefore relation (3.3) has the following geometrical interpretation: to each of vectors $a_j(t)$ assign a curve in $\mathbb{R}^n$. Relation (3.3) implies that all these curves cannot be placed on a linear subspace of dimension less than n.

Let us also remark that a symmetric matrix K is said to be a positively definite, $K > 0$, if the corresponding quadratic form $z^* K z$ is positively definite: $z^* K z > 0 \quad \forall\, z \in \mathbb{R}^n, \ z \neq 0$.

Proof of Theorem 3.1. At first we note that condition 1$'$) is simply another formulation of condition 1). It can easily be seen that condition 2$'$) yields condition 2). Suppose, condition 2$'$) is not satisfied. Then there exists a vector $z \in \mathbb{R}^n$, $z \neq 0$ such that $z^* e^{At} b = 0$, $\forall t \in (\tau_1, \tau_2)$. Since vector function $z^* e^{At} b$ is the analytic one, by the principle of analytic continuation we have an identity $z^* e^{At} b = 0$, $\forall\, t \in \mathbb{R}^1$. Therefore condition 2) is not satisfied and the equivalence of 2) and 2$'$) is proved.

Let us show that from the property of controllability, formula (3.3) follows.

Suppose, formula (3.3) is not valid. Then there exists a nonzero vector $z \in \mathbb{R}^n$ such that $z^* e^{At} b = 0$, $\forall t \in \mathbb{R}^1$.

Recall that the solution $x(t)$ of system (3.1) can be given in the Cauchy form

$$x(t) = e^{At}\Big(x(0) + \int_0^t e^{-A\tau} b\,\xi(\tau)\,d\tau\Big). \tag{3.6}$$

Let a number $T > 0$ be fixed and in the definition of controllability we assume that $x(T) = x_1 = 0$. Then identity (3.6) has the form

$$x_0 + \int_0^T e^{-A\tau} b\,\xi(\tau)\,d\tau = 0. \tag{3.7}$$

Now we multiply both sides of this identity by the vector z^*:

$$z^* x_0 + \int_0^T z^* e^{-A\tau} b\,\xi(\tau)\,d\tau = 0. \tag{3.8}$$

Since $z^* e^{At} b = 0$, $\forall t \in \mathbb{R}^1$, by identity (3.8) we obtain that $z^* x_0 = 0$. However the last relation cannot be true since x_0 is an arbitrary vector from $\mathbb{R}^n$.

Thus, if formula (3.3) is not valid, then system (3.1) is not controllable, i.e., the controllability implies formula (3.3).

Now we show that condition 2) yields condition 1).

Suppose, condition 1) is not true. Then condition 1′) is not also true. Hence there exists a nonzero vector $z \in \mathbb{R}^n$ such that $z^* A^k b = 0$, $\forall k = 0, \ldots, n-1$.

We also prove that $z^* A^n b = 0$. By the Cayley identity we have

$$A^n + \delta_{n-1} A^{n-1} + \ldots + \delta_1 A + \delta_0\, I = 0, \tag{3.9}$$

where δ_j are coefficients of characteristic polynomial of the matrix A. By (3.9)

$$z^* A^n b = -\delta_{n-1} z^* A^{n-1} b - \ldots - \delta_0 z^* b = 0. \tag{3.10}$$

Multiplying both sides of (3.9) by A and using (3.10), we obtain

$$z^* A^{n+1} b = \delta_{n-1} z^* A^n b - \ldots - \delta_0 z^* A b = 0.$$

Repeating this procedure, we obtain that $z^* A^k b = 0$ for any natural k. Whence it follows that

$$z^* e^{At} b = \sum_{k=0}^{\infty} z^* \frac{(At)^k}{k!} b = \sum_{k=0}^{\infty} \frac{z^* A^k b t^k}{k!} = 0.$$

It is obvious that condition 2) is not true.

Thus, condition 2) implies condition 1).

Now we prove that condition 1) results in condition 3).

Assume that 3) is not valid. Hence there exists a nonzero vector $z \in \mathbb{R}^n$ such that

$$z^* K z = \int_{\tau_1}^{\tau_2} z^* e^{-At} b\, b^* e^{At} z\, dt = 0.$$

The last relation can be given in the following form

$$\int_{\tau_1}^{\tau_2} |z^* e^{-At} b|^2 dt = 0.$$

Hence $z^* e^{-At} b = 0, \forall t \in (\tau_1, \tau_2)$. In this case, as it was already noted at the beginning of the proof, we have $z^* e^{At} b = 0$ for all $t \in \mathbb{R}^1$. Differentiating the expression $z^* e^{At} b$ k times, we arrive at the following identities

$$z^* A^k e^{At} b = 0, \qquad \forall t \in \mathbb{R}^1, \quad k = 0, 1, \ldots$$

Putting $t = 0$, we obtain $z^* b = 0,\ z^* Ab = 0, \ldots, z^* A^{n-1} b = 0$. It means that condition 1) is not valid.

Thus, condition 1) implies condition 3).

Let us prove now that from 3) it follows the controllability. For this purpose we choose an input $\xi(t)$ in the form

$$\xi(t) = b^* e^{-A^* t} \xi_0,$$

where the vector ξ_0 will be determined below.

By the Cauchy formula we obtain

$$x(T) = e^{AT} \left(x(0) + \int_0^T e^{-At} b\, b^* e^{-A^* t} \xi_0\, dt \right).$$

The last relation can be written as

$$e^{-AT}x_1 - x_0 = K\xi_0. \tag{3.11}$$

Since $\det K \neq 0$, equation (3.11) is always solvable and for

$$\xi_0 = K^{-1}(e^{-AT}x_1 - x_0),$$

we obtain the required input $\xi(t)$ that in a time T transforms the solution $x(t)$ from the state $x(0) = x_0$ into the state $x(T) = x_1$.

So, we have

$$\textit{controllability} \Rightarrow 2) \Rightarrow 1) \Rightarrow 3) \Rightarrow \textit{controllability}.$$

The proof of theorem is completed.

Consider now several additional important properties of the controllability.

Theorem 3.2. *The following conditions are equivalent and each of them is necessary and sufficient for a controllability of system* (3.1)*:*

4) *a linear space L, spanned by the complex-valued n-vectors to be composed the matrix $(pI - A)^{-1}b$ for $p \in \mathbb{C}$, $p \neq \lambda_j(A)$, coincides with an n-dimensional space $\mathbb{C}^n$:*

$$L\{(pI - A)^{-1}b \mid p \in \mathbb{C}, \ \ p \neq \lambda_j(A)\} = \mathbb{C}^n, \tag{3.12}$$

Here $\lambda_j(A)$ are the eigenvalues of A.

4′) *for any set $\Omega \subset \mathbb{C}$, having a limiting point different from $\lambda_j(A)$, the following relation holds*

$$L\{(pI - A)^{-1}b \mid p \in \Omega\} = \mathbb{C}^n, \tag{3.13}$$

5) *there does not exist a nonsingular $n \times n$-matrix S such that matrices $S^{-1}AS$ and $S^{-1}b$ take the form*

$$S^{-1}AS = \begin{pmatrix} A_{11} & A_{12} \\ 0 & A_{22} \end{pmatrix}, \qquad S^{-1}b = \begin{pmatrix} b_1 \\ 0 \end{pmatrix},$$

6) $\operatorname{rank}(q_0, \ldots, q_{n-1}) = n$, *where q_j are coefficients of the polynomial*

$$q_{n-1}p^{n-1} + \ldots + q_0 = \big(\det(pI - A)\big)(pI - A)^{-1}b,$$

7) *there does not exist a vector $z \neq 0$ such that it satisfies the relations $A^*z = \lambda z$, $z^*b = 0$, where λ is some number.*

Proof. We prove first that condition 4) is equivalent to condition 4′) For this purpose it is sufficient to prove that 4) implies 4′). Assuming the opposite, we observe that there exists a nonzero vector $z \in \mathbb{C}^n$ such that

$$z^*(pI - A)^{-1}b = 0, \qquad \forall\, p \in \Omega.$$

Since $z^*(pI - A)^{-1}b$ is analytic on Ω, by the principle of analytic continuation we obtain the following identity

$$z^*(pI - A)^{-1}b = 0, \qquad \forall\, p \neq \lambda_j(A),$$

i.e., condition 4) is not true. Thus, 4) yields 4′).

Now we prove that 1′) and 4) are the same. Suppose that condition 4′) is not satisfied for $\Omega = \left\{p \middle| \frac{|A|}{|p|} < 1\right\}$. Then there exists a nonzero vector $z \in \mathbb{C}^n$ such that

$$z^*(pI - A)^{-1}b = 0, \qquad \forall\, p \in \Omega. \tag{3.14}$$

Since for $|p| > |A|$ the following expansion holds

$$\frac{1}{p} z^* \left(I - \frac{A}{p}\right)^{-1} b = \frac{1}{p} z^* \left(I + \frac{A}{p} + \frac{A^2}{p^2} + \ldots\right) b, \tag{3.15}$$

by (3.14) and (3.15) we obtain

$$z^*b + \frac{z^*Ab}{p} + \frac{z^*A^2b}{p^2} + \ldots = 0, \qquad \forall\, p \in \Omega. \tag{3.16}$$

This relation can be regarded as an expansion of zero into the Laurent series with the coefficients z^*A^kb. This expansion is unique and we have $z^*A^kb = 0$, $\forall\, k = 0, 1, \ldots$

Thus, if 4) is not satisfied, then 4′) with Ω, introduced above, is not also satisfied and therefore 1′) is not satisfied too.

If 1′) is not satisfied, then, as it was shown by means of the Cayley identity in the proof of Theorem 3.1, the relations $z^*A^kb = 0$ are satisfied not only for $k = 0, \ldots, n-1$ but also for all natural k. Hence identity (3.16) follows. Then from expansion (3.15) we obtain

$$z^*(pI - A)^{-1}b = 0, \qquad \forall\, p \in \Omega.$$

We see that if 1′) is not satisfied, then 4′) is not also satisfied. The equivalence of 1′) and 4) is proved.

Let us prove the equivalence of Properties 1) and 5).

Assume that 5) does not valid. In this case, by introducing the following notation

$$\tilde{b} = S^{-1}b, \qquad \widetilde{A} = S^{-1}AS,$$

we obtain

$$\widetilde{A}^k\tilde{b} = S^{-1}AS\,S^{-1}AS\ldots S^{-1}AS\,S^{-1}b = S^{-1}A^kb.$$

Therefore

$$(\tilde{b}, \widetilde{A}\tilde{b}, \ldots, \widetilde{A}^{n-1}\tilde{b}) = S^{-1}(b, Ab, \ldots, A^{n-1}b). \tag{3.17}$$

From the assumptions imposed on the matrices $\widetilde{A}$ and $\tilde{b}$ it follows directly that a matrix $(\tilde{b}, \widetilde{A}\tilde{b}, \ldots, \widetilde{A}^{n-1}\tilde{b})$ has the following structure

$$(\tilde{b}, \widetilde{A}\tilde{b}, \ldots, \widetilde{A}^{n-1}\tilde{b}) = \begin{pmatrix} Q \\ 0 \end{pmatrix}.$$

Then $\operatorname{rank}(\tilde{b}, \widetilde{A}\tilde{b}, \ldots, \widetilde{A}^{n-1}\tilde{b}) < n$.

Using (3.17) and the nondegeneracy of matrix S, we obtain

$$\operatorname{rank}(\tilde{b}, \widetilde{A}\tilde{b}, \ldots, \widetilde{A}^{n-1}\tilde{b}) = \operatorname{rank}(b, Ab, \ldots, A^{n-1}b).$$

Consequently,

$$\operatorname{rank}(b, Ab, \ldots, A^{n-1}b) < n.$$

So, condition 1) is not satisfied.

We now assume that condition 1) is not satisfied and $\operatorname{rank}(b, Ab, \ldots$ $\ldots, A^{n-1}b) = r < n$. In this case a nonsingular matrix S can be found such that

$$S^{-1}(b, Ab, \ldots, A^{n-1}b) = \begin{pmatrix} Q \\ 0 \end{pmatrix}, \tag{3.18}$$

where Q is an $r \times nm$-matrix and 0 is a null matrix of dimension $(n-r) \times nm$.

Consider matrices

$$\widetilde{A} = S^{-1}AS = \begin{pmatrix} A_{11} & A_{12} \\ A_{21} & A_{22} \end{pmatrix}, \qquad \tilde{b} = S^{-1}b = \begin{pmatrix} b_1 \\ b_2 \end{pmatrix},$$

where S is a nonsingular matrix satisfying relation (3.18), A_{21} is an $(n - r) \times r$-matrix, and b_2 is an $(n - r) \times m$-matrix.

Obviously, in this case relation (3.17) is satisfied. Hence by (3.18) we obtain

$$(\tilde{b}, \widetilde{A}\tilde{b}, \dots, \widetilde{A}^{n-1}\tilde{b}) = \begin{pmatrix} Q \\ 0 \end{pmatrix}. \tag{3.19}$$

Consequently, $b_2 = 0$ and

$$\widetilde{A}\tilde{b} = \begin{pmatrix} A_{11}b_1 \\ A_{21}b_1 \end{pmatrix}.$$

From (3.19) it follows that $A_{21}b_1 = 0$. Hence we obtain

$$\widetilde{A}^2\tilde{b} = \widetilde{A}(\widetilde{A}\tilde{b}) = \begin{pmatrix} A_{11}^2 b_1 \\ A_{21}A_{11}b_1 \end{pmatrix}$$

and by (3.19) $A_{21}A_{11}b_1 = 0$. In this case we have

$$\widetilde{A}^3\tilde{b} = \widetilde{A}(\widetilde{A}^2\tilde{b}) = \begin{pmatrix} A_{11}^3 b_1 \\ A_{21}A_{11}^2 b_1 \end{pmatrix}.$$

Comparing this expression with (3.19), we find that $A_{21}A_{11}^2 b_1 = 0$.

Repeating this procedure, we obtain

$$A_{21}A_{11}^k b_1 = 0, \qquad \forall\, k = 0, 1, \dots, n-2. \tag{3.20}$$

Since A_{11} is an $r \times r$-matrix, by the Cayley identity A_{11}^{n-1} is a linear combination of matrices $A_{11}^{n-2}, \dots, A_{11}, I$. Hence by (3.20)

$$A_{21}A_{11}^{n-1} b_1 = 0. \tag{3.21}$$

Since, as it was proved above, we have

$$Q = (b_1, A_{11}b_1, \dots, A_{11}^{n-1}b_1)$$

and $\operatorname{rank} Q = r$, by (3.20) and (3.21) the matrix A_{21} annuls all vectors $z \in \mathbb{R}^r$:

$$A_{21}\, z = 0, \qquad \forall\, z \in \mathbb{R}^r.$$

Therefore, $A_{21} = 0$.

Finally, we have proved that if 1) is not satisfied, then 5) is not also satisfied.

The proof of equivalence of 1) and 5) is completed.

The equivalence of 4) and 6) is almost obvious. If 6) is not satisfied, then there exists a nonzero vector $z \in \mathbb{C}^n$ such that

$$z^* q_j = 0, \qquad \forall\, j = 0, \ldots, n-1. \tag{3.22}$$

Hence we obtain

$$z^*(pI - A)^{-1} b = 0, \qquad \forall\, p \neq \lambda_j(A). \tag{3.23}$$

If 4) is not satisfied, then there exists a nonzero vector $z \in \mathbb{C}^n$ such that relation (3.23) is valid. In this case we have

$$z^* q_{n-1} p^{n-1} + \ldots + z^* q_0 = 0, \qquad \forall\, p \in \mathbb{C}.$$

Hence (3.22) is true and 6) is not satisfied.

Now we prove the equivalence of 1) and 7).

If there exists a nonzero vector $z \in \mathbb{C}^n$ such that $z^* b = 0$ and $A^* z = \lambda z$, then

$$z^* A^k b = \lambda^k z^* b = 0, \qquad \forall\, k = 1, \ldots$$

Hence 1') is not satisfied.

Suppose that 1) is not satisfied. Denote by L a linear space spanned by column-vectors of the matrix

$$(b, Ab, \ldots, A^{n-1} b).$$

From the Cayley identity it follows that $AL = L$ and from the unsatisfiability of 1) it follows that $\dim L < n$.

Denote by $L^\perp$ a linear subspace orthogonal to L. It is obvious that $\dim L^\perp > 0$, $A^* L^\perp = L^\perp$ and $z^* b = 0$, $\forall\, z \in L^\perp$.

In the sequel we make use of the fact that in an invariant linear subspace there exists at least one eigenvector z, namely

$$A^* z = \lambda z, \qquad z \in L^\perp.$$

From $z^* b = 0$, $\forall\, z \in L^\perp$ it follows that 7) is not satisfied.

The theorem is proved.

3.2 Observability

Consider now the notion of observability [23].

Definition 3.2. *System* (3.1) *is said to be observable if for any number* $T > 0$ *the following relations*

$$\xi(t) = 0, \quad \sigma(t) = 0, \qquad \forall t \in [0, T]$$

imply $x(t) = 0$, $\forall t \in [0, T]$.

Let us clarify this definition.

It is obvious that for $\xi(t) = 0$ on $[0, T]$ we have $x(t) = 0$, $\forall t \in [0, T]$ under the initial data $x(0) = 0$ and therefore $\sigma(t) = c^* x(t) = 0$, $\forall t \in [0, T]$. However for the zero output $\sigma(t)$ the zero solution $x(t)$ is not always unique. For such systems an output $\sigma(t)$ does not determine uniquely the state $x(t)$ and such systems are called nonobservable.

Sometimes for definition of observability the following property is used which is the same as that given in Definition 3.2.

Theorem 3.3. *For system* (3.1) *to be observable, the necessary and sufficient condition is as follows: for* $T > 0$ *relations*

$$\xi_1(t) = \xi_2(t), \quad \sigma_1(t) = \sigma_2(t), \qquad \forall t \in [0, T]$$

imply the identity

$$x_1(t) = x_2(t) \qquad \forall t \in [0, T].$$

Here $x_1(t)$ *is a solution of system* (3.1) *with the vector function* $\xi_1(t)$ *such that* $\sigma_1(t) = c^* x_1(t)$.

Similarly, $x_2(t)$ *is a solution of system* (3.1) *with the vector function* $\xi_2(t)$ *such that* $\sigma_2(t) = c^* x_2(t)$.

Proof. To prove this theorem it is sufficient to show that the property, stated in this theorem, follows from the observability. Subtracting from the system

$$\frac{dx_1}{dt} = Ax_1 + b\xi_1, \qquad \sigma_1 = c^* x_1$$

the system

$$\frac{dx_2}{dt} = Ax_2 + b\xi_2, \qquad \sigma_2 = c^* x_2,$$

we obtain

$$\frac{d(x_1 - x_2)}{dt} = A(x_1 - x_2) + b(\xi_1 - \xi_2),$$

$$(\sigma_1 - \sigma_2) = c^*(x_1 - x_2).$$

Since this system is observable, we can easily see that from the relations

$$\xi_1(t) - \xi_2(t) = 0, \quad \sigma_1(t) - \sigma_2(t) = 0, \qquad \forall t \in [0,T]$$

it follows that

$$x_1(t) - x_2(t) = 0, \qquad \forall t \in [0,T].$$

Then it is clear that the identities

$$\xi_1(t) = \xi_2(t), \quad \sigma_1(t) = \sigma_2(t), \qquad \forall t \in [0,T]$$

imply the following relation

$$x_1(t) = x_2(t), \qquad \forall t \in [0,T],$$

which completes the proof of theorem.

As it was noted above, the property of observability can be stated in the following way: the relations $\xi(t) = 0$, $\sigma(t) = 0$, $\forall t \in [0,T]$ yield the relation $x(0) = 0$.

If we recall that

$$\sigma(t) = c^* e^{At}\left(x(0) + \int_0^t e^{-A\tau} b\,\xi(\tau)\,d\tau\right),$$

then for $\sigma(t) \equiv 0$, $\xi(t) \equiv 0$ we obtain

$$c^* e^{At} x(0) \equiv 0. \tag{3.24}$$

Thus, system (3.1) is observable if and only if from relation (3.24) it follows that $x(0) = 0$.

Rewriting (3.24) in the following form

$$x(0)^* \, e^{A^* t} c \equiv 0$$

and comparing it with condition 2) of controllability of Theorem 3.1, we obtain the following.

Theorem 3.4. *System* (3.1) *is observable if and only if*

$$L\{e^{A^*t}c \mid t \in (-\infty, +\infty)\} = \mathbb{R}^n. \tag{3.25}$$

This result makes it possible to restate properties 1)–7) of the controllability of Theorems 3.1 and 3.2 in terms of observability.

We also remark that from the above-mentioned results it follows that the property of controllability can be stated in terms of the matrices A and b only and the property of observability that in terms of A and c without using system (3.1). Therefore we can consider controllable pairs of matrices (A, b), assuming that property 1) is valid (or any other of equivalent condition 2)–7)). We also consider observable pairs of matrices (A, c), assuming that relation (3.25) is valid (or another condition equivalent to it).

In this case Theorem 3.4 implies the following result.

Theorem 3.5 (The Kalman duality theorem). *A pair* (A, c) *is observable if and only if a pair* (A^*, c) *is controllable.*

Consider system (3.1) in the case that $m = 1$, i.e., b is a vector.

The following criterion for a controllability and observability is important which can be stated in terms of the transfer function.

Theorem 3.6. *System* (3.1) *is observable and controllable if and only if the polynomials* $\det(pI - A)$ *and* $W(p) \det(pI - A)$ *have no common zeros.*

Proof. We assume that polynomials $\det(pI - A)$ and $W(p) \det(pI - A)$ have a common zero $p = p_0$. Then a polynomial

$$q(p) = \det(pI - A)(pI - A)^{-1}b$$

satisfies a relation

$$A\, q(p_0) = p_0\, q(p_0).$$

If a pair (A, b) is not controllable, then the theorem is proved. Now suppose that the controllability exists. Then by virtue of condition 6) of controllability the coefficients q_j of the polynomial $q(p)$ are linear independent vectors and therefore $q(p_0) \neq 0$.

In addition,

$$\begin{aligned} c^* q(p_0) &= \det(p_0 I - A)\, c^*(p_0 I - A)^{-1} b = \\ &= -\det(p_0 I - A)\; W(p_0) = 0. \end{aligned}$$

Thus, there exists a nonzero vector $q(p_0)$ such that

$$A\,q(p_0) = p_0\,q(p_0), \qquad c^*\,q(p_0) = 0.$$

By the Kalman duality theorem and condition 7) of controllability we conclude that a pair (A, c) is not observable.

We show that if system (3.1) is not controllable, then there exists a corresponding common zero $p = p_0$.

Suppose, a pair (A, b) is not controllable. Then by virtue of condition 5) of controllability there exists a nonsingular matrix S such that

$$\widetilde{A} = S^{-1}AS = \begin{pmatrix} A_{11} & A_{12} \\ 0 & A_{22} \end{pmatrix}, \qquad \tilde{b} = S^{-1}b = \begin{pmatrix} b_1 \\ 0 \end{pmatrix}.$$

We have

$$\tilde{c} = S^*c = \begin{pmatrix} c_1 \\ c_2 \end{pmatrix}.$$

Since $W(p)$ is invariant with respect to replacement of A, b, c by $\widetilde{A}$, $\tilde{b}$, $\tilde{c}$ respectively, we obtain

$$W(p) = \tilde{c}^*(\widetilde{A} - pI)^{-1}\tilde{b} = \tilde{c}^* \begin{pmatrix} (A_{11} - pI)^{-1}b_1 \\ 0 \end{pmatrix} =$$

$$= c_1^*(A_{11} - pI)^{-1}b_1.$$

We can easily see that the denominator $c_1^*(A_{11} - pI)^{-1}b_1$ is $\det(pI - A_{11})$ and its degree less than n. Thus the polynomials $\det(pI - A)$ and $\det(pI - A)\,W(p)$ have a common zero.

The case that a pair (A, c) is not observable can be considered in just the same way.

Theorem 3.6 is proved.

We now discuss the form of system (3.1) in the case that it is not controllable. In this case by condition 5) of controllability there exists a matrix S such that

$$S^{-1}AS = \begin{pmatrix} A_{11} & A_{12} \\ 0 & A_{22} \end{pmatrix}, \qquad S^{-1}b = \begin{pmatrix} b_1 \\ 0 \end{pmatrix}. \tag{3.26}$$

Having performed a linear change of variables

$$x = Sy, \qquad y = \begin{pmatrix} y_1 \\ y_2 \end{pmatrix}, \tag{3.27}$$

where the dimension of vector y_1 is equal to the dimension of columns of the matrix b_1, we obtain

$$\begin{aligned} \dot{y} &= S^{-1}AS\,y + S^{-1}b\xi, \\ \sigma &= c^* Sy \end{aligned}$$

and by (3.26) and (3.27)

$$\begin{aligned} \dot{y}_1 &= A_{11}y_1 + A_{12}y_2 + b_1\xi, \\ \dot{y}_2 &= A_{22}y_2. \end{aligned}$$

Thus, in a system, which is not controllable, the subsystem

$$\dot{y}_2 = A_{22}\,y_2$$

can be separated out such that there are no control actions in it.

3.3 A special form of the systems with controllable pair (A, b)

Suppose that b is a vector, i.e., $m = 1$. If the pair (A, b) is assumed to be controllable, then by Theorem 3.1 the vectors

$$\begin{aligned} e_n &= b, \\ e_{n-1} &= (A + a_n I)b, \\ e_{n-2} &= (A^2 + a_n A + a_{n-1}I)b, \\ &\dots\dots\dots\dots\dots\dots \\ e_1 &= (A^{n-1} + a_n A^{n-2} + \ldots + a_2 I)b, \end{aligned} \tag{3.28}$$

where the coefficients a_j being those of the characteristic polynomial $\det(pI - A) = p^n + a_n p^{n-1} + \ldots + a_1$ are linearly independent and form a basis $\mathbb{R}^n$. The matrix A transforms the vectors of this basis in the following way

$$\begin{aligned} Ae_n &= e_{n-1} - a_n e_n, \\ Ae_{n-1} &= e_{n-2} - a_{n-1}e_n, \\ &\dots\dots\dots\dots\dots\dots \\ Ae_1 &= -a_1 e_n. \end{aligned} \tag{3.29}$$

Then for this basis we obtain

$$A = \begin{pmatrix} 0 & 1 & 0 & \dots & 0 \\ 0 & 0 & 1 & \ddots & \vdots \\ \vdots & \vdots & \ddots & \ddots & 0 \\ 0 & 0 & \dots & 0 & 1 \\ -a_1 & -a_2 & \dots & \dots & -a_n \end{pmatrix}, \qquad b = \begin{pmatrix} 0 \\ \vdots \\ \vdots \\ 0 \\ 1 \end{pmatrix} \tag{3.30}$$

Thus, if b and c are vectors and the pair (A, b) is controllable, then the system

$$\frac{dx}{dt} = Ax + b\xi, \quad \sigma = c^* x \tag{3.31}$$

can be given in the following scalar form:

$$\begin{aligned} \dot{x}_1 &= x_2, \\ &\vdots \\ \dot{x}_{n-1} &= x_n, \\ \dot{x}_n &= -a_n x_n - \dots - a_1 x_1 + \xi, \\ \sigma &= c_1 x_1 + \dots + c_n x_n. \end{aligned} \tag{3.32}$$

System (3.32) is equivalent to an equation

$$N\left(\frac{d}{dt}\right)\sigma = M\left(\frac{d}{dt}\right)\xi, \tag{3.33}$$

where the polynomials $N(p)$ and $M(p)$ are the following

$$N(p) = p^n + a_n p^{n-1} + \dots + a_1, \quad M(p) = c_n p^{n-1} + \dots + c_1.$$

By system (3.31) and equation (3.33) the equivalence of description of controllable linear blocks is proved.

3.4 Stabilization. The Nyquist criterion

In Chapter 1 we have discussed various aspects of the phenomenon of instability. This raises a question of whether it is possible to stabilize system (3.1) (i.e., to make it stable) by choice of the input

$$\xi = S^* x,$$

where S is a constant $m \times n$-matrix. For the answer to this question we prove an auxiliary statement.

Let the following matrices, namely an $n \times n$-matrix A, an $m \times m$-matrix D, an $n \times m$-matrix B, and an $m \times n$-matrix C be given.

Lemma 3.1 (The Schur lemma). *If* $\det A \neq 0$, *then*

$$\det \begin{pmatrix} A & B \\ C & D \end{pmatrix} = \det A \ \det(D - CA^{-1}B). \tag{3.34}$$

If $\det D \neq 0$, *then*

$$\det \begin{pmatrix} A & B \\ C & D \end{pmatrix} = \det D \ \det(A - BD^{-1}C). \tag{3.35}$$

Proof. Suppose, $Q = -A^{-1}B$. Then

$$\begin{pmatrix} A & B \\ C & D \end{pmatrix} \begin{pmatrix} I_n & Q \\ 0 & I_m \end{pmatrix} = \begin{pmatrix} A & 0 \\ C & D - CA^{-1}B \end{pmatrix}.$$

Formula (3.34) follows directly from the last relation. The proof of (3.35) is in the same way.

Corollary 3.1. *If K and M are $n \times m$-matrices, then*

$$\det(I_n + KM^*) = \det(I_m + M^*K). \tag{3.36}$$

Proof. By (3.34) we have

$$\det \begin{pmatrix} I_n & K \\ -M^* & I_m \end{pmatrix} = \det I_n \ \det(I_m + M^*K).$$

By (3.35)

$$\det \begin{pmatrix} I_n & K \\ -M^* & I_m \end{pmatrix} = \det I_m \ \det(I_n + KM^*).$$

Formula (3.36) follows from two last relations.

If $\xi = S^*x$, then

$$\frac{dx}{dt} = (A + bS^*)x,$$

$$\sigma = c^*x$$

and by (3.36) the characteristic polynomial of the matrix $A + bS^*$ takes the form

$$\det(pI - A - bS^*) = \det(pI - A)\ \det(I_m - S^*(A - pI)^{-1}b). \tag{3.37}$$

The following theorem states that in the case $m = 1$ for a controllable system, by fitting S the stabilization is always possible. In addition, by choice of S the characteristic polynomial of matrix $A + bS^*$ can be made arbitrary.

Consider now a polynomial of degree n:

$$\psi(p) = p^n + \psi_{n-1}p^{n-1} + \ldots + \psi_0.$$

Theorem 3.7. *If $m = 1$ and a pair (A, b) is controllable, then for any polynomial $\psi(p)$ there exists a vector $S \in \mathbb{R}^n$ such that*

$$\det(pI - A - bS^*) = \psi(p). \tag{3.38}$$

Proof. We recall the notation

$$q(p) = \big(\det(pI - A)\big)\ (pI - A)^{-1}b,$$
$$q(p) = q_{n-1}p^{n-1} + \ldots + q_0.$$

By (3.37)

$$\det(pI - A - bS^*) = \det(pI - A) - S^*q(p).$$

For proving the theorem it is sufficient to show a solvability of the following system

$$\psi_j = \delta_j - S^*q_j, \qquad j = 1, \ldots, n. \tag{3.39}$$

Here δ_j are coefficients of polynomial $\det(pI - A)$:

$$\det(pI - A) = p^n + \delta_{n-1}p^{n-1} + \ldots + \delta_0.$$

The pair (A, b) is controllable. This implies that by condition 6) the vectors q_j are linearly independent and, consequently, the required vector S can be determined as

$$S = \begin{pmatrix} q_0^* \\ \vdots \\ q_{n-1}^* \end{pmatrix}^{-1} \begin{pmatrix} \delta_0 - \psi_0 \\ \vdots \\ \delta_{n-1} - \psi_{n-1} \end{pmatrix}.$$

This concludes the proof of theorem.

In many cases the input ξ is a linear function of output σ

$$\xi = \mu\sigma,$$

where μ is some $m \times l$-matrix (see Fig. 3.1). In this case the matrix S is of a special form

$$S^* = \mu c^*.$$

Further we shall assume that the characteristic polynomial $\delta(p)$ of the matrix A has no zeros on the imaginary axis. The number of its zeros to the right of the imaginary axis is denoted by k_o. The number k_o is called a degree of instability of an open system.

From the point of view of the control theory it is convenient to consider the case $\mu = 0$ as an open loop (see Fig. 3.1).

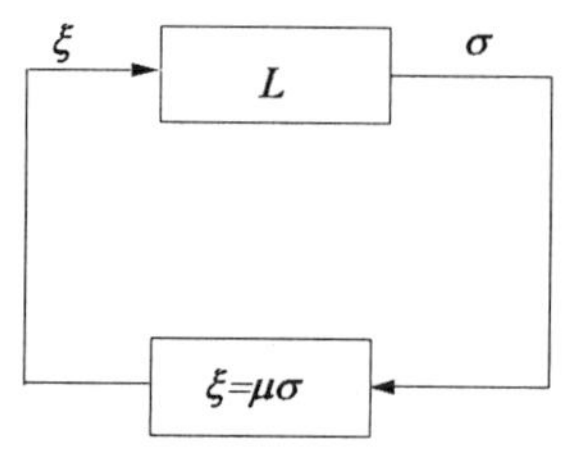

Fig. 3.1

The number of zeros of the polynomial

$$\det(pI - A - b\mu c^*),$$

which are placed to the right of the imaginary axis, is called a degree of instability of the closed system and is denoted by k_c.

Applying the Hermit-Mikhailov formula (see Chapter 1) to both sides of (3.37) being rewritten as

$$\det(pI - A - b\mu c^*) = \delta(p) \det\big(I_m + \mu\, W(p)\big),$$

we obtain

$$\pi(n - 2k_c) = \pi(n - 2k_o) + 2\pi k_1.$$

Here

$$k_1 = \frac{\Delta \text{ Arg } \det(I_m + \mu W(i\omega)) \,|_{-\infty}^{+\infty}}{2\pi}.$$

Note that for the application of the Hermit-Mikhailov formula it is necessary for the following inequality

$$\det(i\omega\, I - A - b\mu c^*) \neq 0, \qquad \forall\, \omega \in \mathbb{R}^1$$

to be satisfied. This inequality is satisfied if we take

$$\det\left(I_m + \mu\, W(i\omega)\right) \neq 0, \qquad \omega \in \mathbb{R}^1. \tag{3.40}$$

Finally, we arrive at the following result.

Theorem 3.8. *Let*

$$\delta(i\omega) \neq 0, \qquad \forall\, \omega \in \mathbb{R}^1 \tag{3.41}$$

and inequality (3.40) *be satisfied. Then the Nyquist formula*

$$k_c = k_o - k_1 \tag{3.42}$$

is valid.

The Nyquist criterion

Under assumptions (3.40) *and* (3.41) *system* (3.1) *is stable if and only if* $k_o = k_1$.

The Nyquist criterion holds much favor due to its geometric interpretation in the case $m = l = 1$.

Recall that in place of the description of block by differential or integral operators the hodograph of its frequency response (Nyquist plot) on the complex plane is often given (Fig. 3.2).

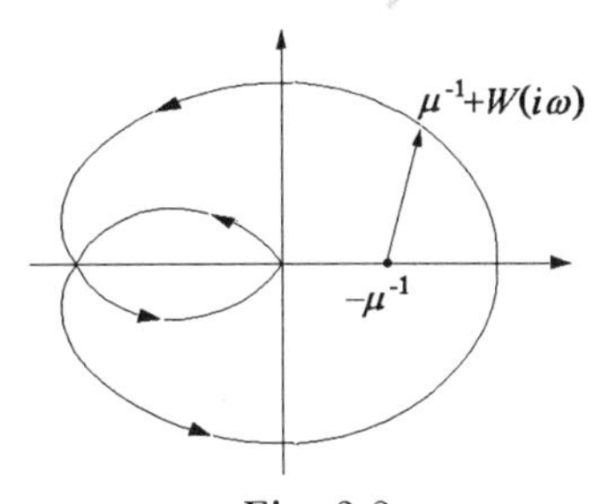

Fig. 3.2

In the case that $m = l - 1$ the number k_1 can be given by

$$k_1 = \frac{\Delta \operatorname{Arg}(\mu^{-1} + W(i\omega))\big|_{-\infty}^{+\infty}}{2\pi}.$$

From this relation and Fig. 3.2 we can conclude that k_1 is the number of counterclockwise rounds of a vector such that its origin is the point $-\mu^{-1}$ and the end is at the point on the hodograph. Here ω running from $-\infty$ to $+\infty$.

In the case, shown in Fig. 3.2, we have $k_1 = 1$.

Since k_o can be obtained by the Hermit-Mikhailov formula, the Nyquist criterion is effective for stabilization of linear systems.

In conclusion we note that definitions of controllability and observability can also be introduced for more general systems than (3.1). In the nonlinear generalizations of these notions, in place of mathematical tools, operating on linear subspaces, some smooth manifolds can be introduced [20, 35] in the natural way. The Nyquist criterion is a part of the linear theory. The latter can be found in many books on the control theory, for example, in [11, 12, 18, 34, 39].

3.5 The time-varying stabilization. The Brockett problem

In 1999 the book [36] has published in which many interesting and important problems of the control theory have been formulated.

In particular, R. Brockett has stated the following problem.

"Given a triple constant matrices (A, b, c^*) under what circumstances does there exist a time-dependent matrix $K(t)$ such that the system

$$\frac{dx}{dt} = Ax + bK(t)c^*x, \quad x \in \mathbb{R}^n \tag{3.43}$$

is asymptotically stable ?"

Note that for system (3.43) the problem of stabilization due to a constant matrix K is a classic problem in the control theory (see §3.4). From this point of view the Brockett problem can be restated in the following way.

How much does the introduction of matrices $K(t)$, depending on time t, enlarge the possibilities of a classic stabilization ?

In studying the stabilization problems of mechanical systems it is sometimes necessary to consider another class of stabilizing matrices $K(t)$. These matrices must be periodic and have a zero mean value on the period $[0, T]$:

$$\int_0^T K(t)\,dt = 0. \tag{3.44}$$

Consider, for example, a linear approximation in a neighborhood of the equilibrium position of a pendulum with vertically oscillating pendulum pin

$$\ddot{\theta} + \alpha\dot{\theta} + (K(t) - \omega_o^2)\theta = 0, \tag{3.45}$$

where α and ω_o are positive numbers.

Here the functions $K(t)$ being most frequently considered are of the form [15, 32] $\beta \sin \omega t$ and those of the following form [3, 4]

$$K(t) = \begin{cases} \beta & \text{for } t \in [0, T/2) \\ -\beta & \text{for } t \in [T/2, T) \end{cases} \tag{3.46}$$

For such functions $K(t)$ the effect of stabilization of the upper equilibrium position for large ω and small T is well known.

Here the algorithms of constructing the periodic piecewise constant functions $K(t)$, which make it possible to solve in some cases the Brockett problem and stabilization problem in the classes of functions satisfying condition (3.44), are obtained. For pendulum equation (3.45) with $K(t)$ of the form (3.46) the possibility of low-frequency stabilization is proved.

Suppose, there exist matrices K_j $(j = 1, 2)$ such that the systems

$$\frac{dx}{dt} = (A + bK_jc^*)x, \quad x \in \mathbb{R}^n \tag{3.47}$$

have stable linear manifolds L_j and invariant linear manifolds M_j. We assume that $M_j \cap L_j = \{0\}$, $\dim M_j + \dim L_j = n$, and for the positive numbers λ_j, $æ_j$, α_j, β_j the inequalities hold

$$|x(t)| \leq \alpha_j e^{-\lambda_j t}|x(0)|, \quad \forall\, x(0) \in L_j, \tag{3.48}$$

$$|x(t)| \leq \beta_j e^{æ_j t}|x(0)|, \quad \forall\, x(0) \in M_j. \tag{3.49}$$

Suppose also that there exists a matrix $U(t)$ and a number $\tau > 0$ such that for the system

$$\frac{dy}{dt} = (A + bU(t)c^*)y \tag{3.50}$$

the following inclusion

$$Y(\tau)M_1 \subset L_2 \tag{3.51}$$

is valid. Here $Y(t)$ is a fundamental matrix of system (3.50), $Y(0) = I$.

Theorem 3.9. *If the inequality holds*

$$\lambda_1\lambda_2 > æ_1æ_2, \tag{3.52}$$

then there exists a periodic matrix $K(t)$ such that system (3.43) is asymptotically stable.

Proof. Condition (3.52) implies the existence of positive numbers t_1 and t_2 satisfying the following inequalities

$$-\lambda_1 t_1 + æ_2 t_2 < -T, \quad -\lambda_2 t_2 + æ_1 t_1 < -T. \tag{3.53}$$

Here T is an arbitrary positive number.

We now define a periodic matrix $K(t)$ in the following way:

$$\begin{aligned} K(t) &= K_1, \quad \forall t \in [0, t_1), \\ K(t) &= U(t - t_1), \quad \forall t \in [t_1, t_1 + \tau), \\ K(t) &= K_2, \quad \forall t \in [t_1 + \tau, t_1 + t_2 + \tau). \end{aligned} \tag{3.54}$$

The period of the matrix $K(t)$ is as follows: $t_1 + t_2 + \tau$. We show that for sufficiently large T system (3.43) with such a matrix $K(t)$ is asymptotically stable. For this purpose, consider nonsingular matrices S_j, reducing systems (3.47) to the canonical form:

$$\begin{aligned} \frac{dz_j}{dt} &= Q_j z_j, \qquad \dim z_j = \dim L_j, \\ \frac{dw_j}{dt} &= P_j w_j, \qquad \dim w_j = \dim M_j. \end{aligned} \tag{3.55}$$

Here

$$S_j x = \begin{pmatrix} z_j \\ w_j \end{pmatrix}, \tag{3.56}$$

and without loss of generality we may assume that

$$\begin{aligned} |z_j(t)| &\le e^{-\lambda_j t}|z_j(0)|, \\ |w_j(t)| &\le e^{æ_j t}|w_j(0)|. \end{aligned} \tag{3.57}$$

From relations (3.54)–(3.56) it follows that

$$\begin{pmatrix} z_2(t_1+\tau) \\ w_2(t_1+\tau) \end{pmatrix} = S_2 Y(\tau) S_1^{-1} \begin{pmatrix} z_1(t_1) \\ w_1(t_1) \end{pmatrix}. \tag{3.58}$$

Inclusion (3.51) implies that the matrix $S_2Y(\tau)S_1^{-1}$ has the form

$$S_2Y(\tau)S_1^{-1} = \begin{pmatrix} R_{11}(\tau) & R_{12}(\tau) \\ R_{21}(\tau) & 0 \end{pmatrix}.$$

Therefore (3.53) and (3.57) result in the estimates

$$\begin{aligned} |z_2(t_1+t_2+\tau)| &\le |R_{11}(\tau)|e^{-2T}|z_1(0)| + |R_{12}(\tau)|e^{-T}|w_1(0)|, \\ |w_2(t_1+t_2+\tau)| &\le |R_{21}(\tau)|e^{-T}|z_1(0)|. \end{aligned}$$

Whence it follows that for sufficiently large values of T for the initial values from the ball $|x(0)| \le 1$ we have $|x(t_1+t_2+\tau)| \le 1/2$. This relation and the periodicity of the matrix $K(t)$ imply the asymptotic stability of system (3.43).

Suppose now that the matrix $K(t)$ in equation (3.43) is a scalar function, $K_1 = K_2 = K_0$, $\lambda_1 = \lambda_2 = \lambda$, $æ_1 = æ_2 = æ$, $U(t) \equiv U_0$, $K_0U_0 < 0$, the function $|Y(t)|$ is uniformly bounded on the interval $(0, +\infty)$, and there exists a sequence $\tau_j \to +\infty$ such that

$$Y(\tau_j)M_1 \subset L_2. \tag{3.59}$$

Theorem 3.10. *Suppose, an inequality $\lambda > æ$ is satisfied. Then there exists a T-periodic function $K(t)$ such that* (3.44) *is satisfied and system* (3.43) *is asymptotically stable.*

For Theorem 3.10 to be proved it is sufficient to introduce the following function $K(t)$:

$$\begin{aligned} K(t) &= K_0, \ \forall t \in [0, |U_0\tau_j/2K_0|), \\ K(t) &= U_0, \ \forall t \in [|U_0\tau_j/2K_0|, \tau_j + |U_0\tau_j/2K_0|), \\ K(t) &= K_0, \ \forall t \in [\tau_j + |U_0\tau_j/2K_0|, \tau_j + |U_0\tau_j/K_0|). \end{aligned} \tag{3.60}$$

The period of $K(t)$ is equal to $T = \tau_j(1 + |U_0/K_0|)$.

Here τ_j is a sufficiently large number satisfying condition (3.59). The further proof of Theorem 3.10 is just in the same way as the proof of Theorem 3.9.

We now apply Theorem 3.10 to equation (3.45) with the function $K(t)$ of the form (3.46).

Let

$$\alpha^2 < 4(\beta - \omega_0^2). \tag{3.61}$$

Without loss of generality we may assume that $\beta - \omega_0^2 - \alpha^2/4 = 1$.

Put $K_0 = -\beta$, $U_0 = \beta$. From condition (3.61) it follows that the characteristic polynomial of equation (3.45) with $K(t) = U_0$ has complex roots, and, consequently, condition (3.59) with some $\tau_1 > 0$ is satisfied. Then we have $\tau_j = \tau_1 + 2j\pi$.

It can easy be seen that the negativity of real parts of the roots of the characteristic polynomial implies the uniform boundedness of $|Y(t)|$ on $(0, +\infty)$.

For $K(t) = K_0 = -\beta$ the values of λ and æ can be calculated:

$$\lambda = \frac{\alpha}{2} + \sqrt{\frac{\alpha^2}{4} + (\beta + \omega_0^2)}\ ,$$

$$æ = -\frac{\alpha}{2} + \sqrt{\frac{\alpha^2}{4} + (\beta + \omega_0^2)}\ .$$

Thus, all the assumptions of Theorem 3.10 hold, and equation (3.45) with $K(t)$ of the form (3.60) is asymptotically stable for sufficiently large j. This result can be stated in the following way.

Proposition 3.2. *If* (3.61) *is satisfied, then for any number τ there exists a number $T > \tau$ such that equation* (3.45) *with the function $K(t)$ of the form* (3.46) *is asymptotically stable.*

In particular, we see that the stabilization of the upper position of pendulum is possible under the low-frequency vertical oscillation of the pendulum pin. It is natural that in this case the oscillation amplitude a turns out to be large $a = lT^2\beta/8$, where l is a pendulum length, β is the absolute acceleration value divided by l.

For high-frequency oscillation (for small T) the effect of stabilization is well known [3, 4].

For the effective testing of condition (3.51) sometimes the following lemmas are useful.

Consider a system

$$\dot{z} = Qz, \quad z \in \mathbb{R}^n, \tag{3.62}$$

where Q is a constant nonsingular $n \times n$-matrix, and a vector $h \in \mathbb{R}^n$.

Lemma 3.2. *Suppose, a solution $z(t)$ of system (3.62) has the form $z(t) = v(t)+w(t)$, where $v(t)$ is a periodic vector function such that $h^*v(t) \not\equiv 0$, $w(t)$ is a vector function such that*

$$\int_0^{+\infty} |w(\tau)|\, d\tau < +\infty, \qquad \lim_{t\to+\infty} w(t) = 0.$$

Then there exist numbers τ_1 and τ_2 such that the following inequalities

$$h^*z(\tau_1) > 0, \quad h^*z(\tau_2) < 0 \tag{3.63}$$

are satisfied.

Proof. Assuming the opposite, we obtain $h^*z(t) \geq 0$, $\forall t \geq 0$ or $h^*z(t) \leq 0$, $\forall t \geq 0$. Let, for definiteness, $h^*z(t) \geq 0$, $\forall t \geq 0$. The inequality $h^*v(t) \not\equiv 0$ implies that

$$\lim_{t\to+\infty} \int_0^t h^*z(\tau)\, d\tau = +\infty. \tag{3.64}$$

On the other hand, we have

$$\int_0^t h^*z(\tau)\, d\tau = h^*Q^{-1}\big(z(t) - z(0)\big).$$

From the above and the uniform boundedness of $z(t)$ on $(0, +\infty)$ we have a uniform boundedness of the following function

$$\int_0^t h^*z(\tau)\, d\tau,$$

which contradicts relation (3.64). This contradiction proofs the lemma.

Lemma 3.3. *Let $n = 2$ and a matrix Q have complex eigenvalues. Then for any pair of nonzero vectors $h \in \mathbb{R}^2$, $u \in \mathbb{R}^2$ there exist numbers τ_1 and τ_2 such that*

$$h^*e^{Q\tau_1}u > 0, \quad h^*e^{Q\tau_2}u < 0. \tag{3.65}$$

This obvious statement may be regarded as a corollary of Lemma 3.2.

Lemma 3.4. *Let a matrix Q have two complex eigenvalues $\lambda_0 \pm i\omega_0$ and other its eigenvalues $\lambda_j(Q)$ satisfy the condition $\operatorname{Re}\lambda_j(Q) < \lambda_0$.*

Let also for the vectors $h \in \mathbb{R}^n$, $u \in \mathbb{R}^n$ the following inequalities hold

$$\det(h, Q^*h, \ldots, (Q^*)^{n-1}h) \neq 0, \tag{3.66}$$

$$\det(u, Qu, \ldots, Q^{n-1}u) \neq 0. \tag{3.67}$$

Then there exist numbers τ_1 and τ_2 such that

$$h^*e^{Q\tau_1}u > 0, \quad h^*e^{Q\tau_2}u < 0. \tag{3.68}$$

Recall that conditions (3.67) and (3.66) are the conditions of controllability for a pair (Q, u) and those of observability for a pair (Q, h).

To prove Lemma 3.4 it is sufficient to note that the solution $z(t) = e^{Qt}u$ can be represented as

$$z(t) = e^{\lambda_0 t}\big(v(t) + w(t)\big),$$

where $v(t)$ and $w(t)$ satisfy the assumptions of Lemma 3.2. The inequality $h^*v(t) \not\equiv 0$ follows from the observability of (Q, h) and controllability of (Q, u).

Theorem 3.9 and Lemma 3.3 imply the following proposition.

Theorem 3.11. *Let $n = 2$ and there exist matrices K_0 and U_0, satisfying the following conditions:*

1) $\det bK_0c^* = 0$ *and* $Tr\, bK_0c^* \neq 0$,

2) *the matrix* $A + bU_0c^*$ *has complex eigenvalues.*

Then there exists a periodic matrix $K(t)$ of the form (3.54) *such that system* (3.43) *is asymptotically stable.*

For proving this theorem it is sufficient to put $K_1 = K_2 = \mu K_0$, where $|\mu|$ is a sufficiently large number and $\operatorname{Tr}\mu bK_0c^* < 0$. It is obvious that in this case all the assumptions of Theorem 3.9 are true.

Consider now the case that b and c are vectors, $K(t)$ is a piecewise continuous function: $\mathbb{R}^1 \to \mathbb{R}^1$.

Consider a transfer function of system (3.43) $W(p) = c^*(A - pI)^{-1}b$, where p is a complex variable. We assume that $W(p)$ is a nondegenerate function. This means that the pair (A, b) is controllable and the pair (A, c) is observable.

Lemma 3.5. *If a hyperplane $\{h^* z = 0\}$ is an invariant manifold for the system*

$$\dot{x} = (A + \mu bc^*)x, \qquad \mu \neq 0, \tag{3.69}$$

then a pair (A, h) is observable.

Proof. Suppose that (A, h) is not observable. In this case there exist both the vector q and the number γ such that $h^* q = 0, \quad Aq = \gamma q, \quad q \neq 0$ (see Theorems 3.5 and 3.2). The observability of a pair (A, c) implies the inequality $c^* q \neq 0$.

From the invariance of $\{h^* z = 0\}$ with respect to equation (3.69) it follows that for all $z \in \{h^* z = 0\}$ we have $h^*(A + \mu bc^*)^k z = 0$, $k = 1, 2, \ldots$ For $z = q$ and $k = 1$ we obtain $h^* bc^* q = 0$. Hence $h^* b = 0$. For $z = q$ and $k = 2$, using the previous relation, we obtain $h^* Ab = 0$. Continuing in the same way, we obtain further the equalities $h^* A^{k-1} b = 0$. The controllability of the pair (A, b) implies that $h = 0$. Hence the assumption on the unobservability of the pair (A, h) is incorrect. The lemma is proved.

Lemma 3.6. *If the straight line $\{\alpha u\}$, $u \in \mathbb{R}^n$, $\alpha \in \mathbb{R}^1$ is invariant with respect to system* (3.69), *then the pair (A, u) is controllable.*

Proof. The invariance implies that $(A + \mu bc^*)^k u = \gamma_k u$, where $k = 0, 1, \ldots, \ \gamma_k$ are some numbers. From the observability of (A, c) it follows that $c^* u \neq 0$. Therefore for vectors $z \in \mathbb{R}^n$ such that $z^* u = 0$, $z^* Au = 0, \ \ldots, z^* A^{n-1} u = 0$ we obtain the following relations $z^* b = 0$, $z^* Ab = 0, \ldots, z^* A^{n-1} b = 0$. From the controllability of the pair (A, b) it follows that $z = 0$.

Thus, the relations $z^* u = 0, \ldots, z^* A^{n-1} u = 0$ imply that $z = 0$. This proves the controllability of the pair (A, u).

From Theorem 3.9 and Lemmas 3.4–3.6 the following result can be obtained:

Theorem 3.12. *Let $b \in \mathbb{R}^n$, $c \in \mathbb{R}^n$,* $\dim M_1 = 1$, $\dim L_2 = n - 1$, *inequality* (3.52) *be satisfied, for some number $U_0 \neq K_j$ a matrix $A + U_0 bc^* i$ has complex eigenvalues $\lambda_0 \pm i\omega_0$ and its other eigenvalues λ_j satisfy a condition* $\mathrm{Re}\, \lambda_j < \lambda_0$.

Then there exists a periodic function $K(t)$ such that system (3.43) *is asymptotically stable.*

To prove Theorem 3.12 it is sufficient to note that by Lemma 3.5 the inequality $U_0 \neq K_j$, the controllability (A, b), and the observability (A, c) imply the observability of the pair $(A + U_0 bc^*, h)$, where h is a normal vector of the subspace L_2. By Lemma 3.6 the pair $(A + U_0bc^*, u)$ is controllable. Here $u \neq 0$, $u \in M_1$. By Lemma 3.4 there exists number τ such that

$$h^* \exp\left[(A + U_0bc^*)\tau\right]u = 0.$$

Whence it follows that condition (3.51) is satisfied.

Lemmas 3.2–3.6 have been stated for testing inclusion (3.51) in the case that the rotation of the subspace M_1 is caused by the presence of complex eigenvalues. Another approach is a pulse action $U(t) = \mu$ with a large $|\mu|$ in a small time interval. In this case a velocity vector $\dot{x}$ is often close to a vector γb, where γ is a number.

We discuss this approach in more detail.

Consider system (3.69) with large parameter μ: $|\mu| \gg 1$.

Lemma 3.7. *Let $c^*b = 0$ and for the vectors h, u the inequalities $h^*b \neq 0$, $c^*u \neq 0$ be satisfied. Then there exist numbers μ and $\tau(\mu) > 0$ such that $h^*x(\tau, u) = 0$, $\lim\limits_{\mu\to\infty} \tau(\mu) = 0$.*

Proof. Consider the following numbers

$$t_0 = -\frac{h^*u}{\mu h^*bc^*u},$$

$$R = \frac{(1 + 2|\mu||b||c|t_0)|u|}{1 - (2|A|t_0 + 4|\mu||A||b||c|t_0^2)}.$$

We choose the number μ such that $t_0 > 0$ and

$$2|A|t_0 + 4|\mu||A||b||c|t_0^2 < 1.$$

It is obvious that for $t \in [0, 2t_0]$ we have

$$|c^*\dot{x}(t,u)| = |c^*Ax(t,u)| \leq |A||c| \max_{t\in[0,2t_0]} |x(t,u)|.$$

Therefore for $t \in [0, 2t_0]$ we obtain

$$|c^*x(t,u) - c^*u| \leq 2|A||c|t_0 \max_{t\in[0,2t_0]} |x(t,u)|.$$

From equation (3.69) it follows that

$$\begin{aligned}&|x(t,u) - u - \mu bc^*ut| \leq \\ &\leq (2|A|t_0 + 4|\mu||A||b||c|t_0^2) \max_{t\in[0,2t_0]} |x(t,u)|.\end{aligned}$$

The estimation implies the following inequalities

$$|x(t,u)| \leq R, \quad \forall t \in [0, 2t_0],$$

$$|h^*x(t,u) - h^*u - \mu h^*bc^*ut| \leq$$
$$\leq (2|A|t_0 + 4|\mu||A||b||c|t_0^2)R|h|, \quad \forall t \in [0, 2t_0].$$

We can easily see that

$$(2|A|t_0 + 4|\mu||A||b||c|t_0^2)R|h| = O\left(\frac{1}{\mu}\right).$$

Then for large $|\mu|$ there exists $\tau \in [0, 2t_0]$ such that $h^*x(\tau, u) = 0$.

Lemma 3.8. *Let $c^*b \neq 0$ and for the vectors h and u the inequalities $h^*b \neq 0$, $c^*u \neq 0$*

$$\frac{h^*uc^*b}{h^*bc^*u} < 1$$

be satisfied. Then there exist numbers μ and $\tau(\mu) > 0$ such that $h^*x(\tau, u) = 0$, $\lim_{\mu \to \infty} \tau(\mu) = 0$.

Proof. Consider the numbers

$$t_0 = \frac{1}{\mu c^*b} \ln\left(1 - \frac{h^*uc^*b}{h^*bc^*u}\right),$$

$$R = \frac{(1 + |b||c||c^*b|^{-1}(e^{2\mu c^*bt_0} + 1))|u|}{1 - (2|A|t_0 + 2|A||b||c|t_0|c^*b|^{-1}(e^{2\mu c^*bt_0} + 1))}.$$

We choose the number μ such that $t_0 > 0$ and

$$2|A|t_0 + 2|A||b||c|t_0|c^*b|^{-1}(e^{2\mu c^*bt_0} + 1) < 1.$$

For $t \in [0, 2t_0]$ we have

$$|c^*\dot{x}(t,u) - \mu c^*bc^*x(t,u)| \leq |A||c| \max_{t \in [0, 2t_0]} |x(t,u)|.$$

Then for $t \in [0, 2t_0]$ we obtain

$$|c^*x(t,u) - e^{\mu c^*bt}c^*u| \leq \frac{1 - e^{2\mu c^*bt_0}}{-\mu c^*b}|A||c| \max_{t \in [0, 2t_0]} |x(t,u)|.$$

Then from (3.69) it follows that

$$\left|x(t,u) - u - \tfrac{bc^*u}{c^*b}\left(e^{\mu c^* bt} - 1\right)\right| \leq$$

$$\leq \left(2|A|t_0 + 2|A||b||c|t_0|c^*b|^{-1}(e^{2\mu c^* bt_0} + 1)\right) \max_{t\in[0,2t_0]} |x(t,u)|.$$

The estimation results in the following inequalities

$$|x(t,u)| \leq R, \quad \forall t \in [0, 2t_0],$$

$$\left|h^*x(t,u) - h^*u - \frac{h^*bc^*u}{c^*b}\left(e^{\mu c^* bt} - 1\right)\right| \leq$$

$$\leq \left(2|A|t_0 + 2|A||b||c|t_0|c^*b|^{-1}(e^{2\mu c^* bt} + 1)\right) R|h|, \ \forall t \in [0, 2t_0].$$

Obviously

$$(2|A|t_0 + 2|A||b||c|t_0|c^*b|^{-1}(e^{2\mu c^* bt_0} + 1))R|h| = O\left(\frac{1}{\mu}\right).$$

Therefore for large $|\mu|$ there exists $\tau \in [0, 2t_0]$ such that $h^*x(\tau, u) = 0$. The proof of lemma is completed.

Theorem 3.13. *Suppose,* $b \in \mathbb{R}^n$, $c \in \mathbb{R}^n$, $c^*b = 0$, $\dim M_1 = 1$, $\dim L_2 = n - 1$ *and inequality* (3.52) *is valid.*

Then there exists a periodic function $K(t)$ *such that* (3.43) *is asymptotically stable.*

For the proof of theorem it is sufficient to remark that from a controllability of (A, b) and an observability of (A, c) by Lemmas 3.5 and 3.6 it follows an observability of $((A + Kbc^*), h)$ for any $K \neq K_2$ and a controllability of $((A + Kbc^*), u)$ for any $K \neq K_1$. Here h is a vector normal to hyperplane L_2 and u is an nonzero vector from the subspace M_1. Whence it follows that $h^*b \neq 0$ and $c^*u \neq 0$. Therefore by Lemma 3.7 there exist numbers μ and $\tau(\mu)$ such that for $U(t) \equiv \mu$ for system (3.50), condition (3.51) is satisfied.

Theorem 3.14. *Suppose,* $c^*b \neq 0$, *the matrix* A *has a positive eigenvalue* æ *and* $(n-1)$ *eigenvalues with the real parts to be less than* $-\lambda$, $\lambda >$ æ. *We also assume that the inequality holds*

$$\frac{c^*b}{\lim\limits_{p\to æ}(æ - p)W(p)} < 1.$$

Then there exists a periodic function $K(t)$ such that the system (3.43) is asymptotically stable.

Proof. Without loss of generality it can be assumed that the matrix A and vectors b and c take the form

$$A = \begin{pmatrix} æ & 0 \\ 0 & A_2 \end{pmatrix}, \quad b = \begin{pmatrix} b_1 \\ b_2 \end{pmatrix}, \quad c = \begin{pmatrix} c_1 \\ c_2 \end{pmatrix},$$

where A_2 is an $(n-1) \times (n-1)$-matrix, $b_2 \in \mathbb{R}^{n-1}$, $c_2 \in \mathbb{R}^{n-1}$. In this case the vector h normal to subspace $L_1 = L_2$ and the vector $u \in M_1 = M_2$ take the form

$$h = \begin{pmatrix} 1 \\ 0 \end{pmatrix}, \quad u = \begin{pmatrix} 1 \\ 0 \end{pmatrix}.$$

By Lemma 3.8 there exist numbers μ and $\tau(\mu)$ such that for $U(t) \equiv \mu$ for system (3.50) condition (3.51) is satisfied in the case that

$$\frac{c^*b}{c_1 b_1} < 1.$$

It is obvious that $c_1 b_1 \neq 0$ by virtue of the controllability of (A, b) and the observability of (A, c) and in addition we have

$$c_1 b_1 = \lim_{p \to æ} (æ - p) W(p).$$

Thus all the assumptions of Theorem 3.9 are valid and therefore system (3.43) is stabilizable.

Lemma 3.9. *Let an $(n-2)$-dimensional linear subspace L invariant with respect to system (3.69) be placed in the hyperplane $\{h^*z = 0\}$. If the relations $h^*b = 0$, $c^*b = 0$ are satisfied, then a pair (A, h) is observable.*

Proof. From a controllability of a pair (A, b) and from an invariance of L with respect to system (3.69) the relation $b \overline{\in} L$ follows. Then we see that the linear subspace, spanned over b and L, coincides with the hyperplane $\{h^*z = 0\}$. Let us assume that a pair (A, h) is not observable. In this case (see Theorems 3.5 and 3.2) there exist both the vector $q \neq 0$ and the number γ such that $h^*q = 0$, $\quad Aq = \gamma q$. The observability of a pair (A, c) implies the inequality $c^*q \neq 0$. From the invariance of L it follows that $h^*(A + \mu bc^*)^k z = 0$, $k = 0, 1, \ldots$, $z \in L$. Arguing as above, we conclude that for q there exist both the number ν and the vector $z \in L$ such that $q = z + \nu b$. Put $\nu \neq 0$. For $k = 1$ we obtain $\nu h^*Ab = 0$. Then $h^*Ab = 0$. Taking into account last relation and $h^*b = 0$, $c^*b = 0$, for $k = 2$ we obtain that $\nu h^* A^2 b = 0$. Continuing in the same way, we see that for $k = 3, \ldots$ the relation $h^* A^k b = 0$ is valid. By virtue of a controllability of (A, b) we obtain that $h = 0$, which contradicts the definition of vector h.

Thus, for $\nu \neq 0$ the observability of (A, h) is proved. For $\nu = 0$ we have the same line of reasoning as in proving Lemma 3.5.

Theorem 3.15. *Let $b \in \mathbb{R}^n$, $c \in \mathbb{R}^n$, $c^* b = 0$, $\dim M_1 = 1$, $\dim L_2 = n - 2$, the inequality (3.52) be satisfied and for a certain number $U_0 \neq K_j$ the assumption of Theorem 3.12 be valid.*

Then there exists a periodic function $K(t)$ such that system (3.43) is asymptotically stable.

Proof. Note that the integral manifold $\Omega(\mu)$, consisting of trajectories $x(t, x_0)$ of system (3.69) with initial data $x_0 \in L_2$, tends to the hyperplane $\{h^* x = 0\}$ as $\mu \to \infty$. Here h is a vector normal to linear subspace, spanned over L_2 and b. This convergence is the same as described in proving Lemma 3.7.

From Lemma 3.9 it follows that a pair $((A + U_0 bc^*), h)$ is observable. From Lemma 3.6 it follows that a pair $((A + U_0 bc^*), u)$ such that $u \in M_1$, $u \neq 0$ is controllable. Then by Lemma 3.4 we obtain that for system (3.50) with $U(t) = U_0$ the function $h^* y(t, u)$ changes its sign for some values of t. Hence for sufficiently large $|\mu|$ there exists a number $\tau_0(\mu) > 0$ such that $y(\tau_0(\mu), u) \in \Omega(\mu)$.

By using small perturbations of the right-hand side of system (3.43), one can obtain the inequality $c^* y(\tau_0(\mu), u) \neq 0$ (under a small perturbation of the right-hand sides of periodic systems the asymptotic stability is preserved).

Further, letting in (3.50) $U(t) = \mu$ (or $U(t) = -\mu$) on $(\tau_0, \tau]$, we can reach a set L_2, at time $t = \tau$. The sign of μ we choose in such a way that $\tau > \tau_0$ (see the proof of Lemma 3.7).

We see that $y(\tau, u) \in L_2$ and condition (3.51) is satisfied.

Thus, all the assumptions of Theorem 3.9 are valid.

In the case that the transfer function $W(p)$ of system (3.43) is nondegenerate we have

$$W(p) = c^* (A - pI)^{-1} b = \frac{c_n p^{n-1} + \ldots + c_1}{p^n + a_n p^{n-1} + \ldots + a_1},$$

where c_j and a_j are real numbers, and system (3.43) may be written in the following scalar form (see § 3.3):

$$\begin{aligned}
&\dot{x}_1 = x_2, \\
&\dots\dots\dots\dots\dots\dots\dots\dots \\
&\dot{x}_{n-1} = x_n, \\
&\dot{x}_n = -(a_n x_n + \ldots + a_1 x_1) - \\
&-K(t)(c_n x_n + \ldots + c_1 x_1).
\end{aligned} \tag{3.70}$$

Then

$$c = \begin{pmatrix} c_1 \\ \vdots \\ c_n \end{pmatrix}, \qquad x = \begin{pmatrix} x_1 \\ \vdots \\ x_n \end{pmatrix},$$

$$A = \begin{pmatrix} 0 & 1 & 0 & \ldots & 0 \\ 0 & 0 & 1 & \ddots & \vdots \\ \vdots & \vdots & \ddots & \ddots & 0 \\ 0 & 0 & \ldots & 0 & 1 \\ -a_1 & -a_2 & \ldots & \ldots & -a_n \end{pmatrix}, \quad b = \begin{pmatrix} 0 \\ \vdots \\ 0 \\ -1 \end{pmatrix}.$$

Recall that the nondegeneracy of the transfer function $W(p)$ indicates that the following polynomials $c_n p^{n-1} + \ldots + c_1, \quad p^n + a_n p^{n-1} + \ldots + a_1$ have no common zeros.

Let $c_n \neq 0$. We can assume, without loss of generality, that $c_n = 1$.

Theorem 3.16. *Suppose, the following conditions hold:*

1) *for* $n > 2 \quad c_1 \leq 0, \ldots, c_{n-2} \leq 0$,

2)

$$\begin{aligned}
&c_1(a_n - c_{n-1}) > a_1, \\
&c_1 + (a_n - c_{n-1})c_2 > a_2, \\
&\dots\dots\dots\dots\dots\dots\dots\dots \\
&c_{n-2} + (a_n - c_{n-1})c_{n-1} > a_{n-1}.
\end{aligned}$$

Then there does not exist a function $K(t)$ *such that system* (3.43) *is asymptotically stable.*

Proof. Consider a set

$$\Omega = \{x_1 \geq 0, \ldots, x_{n-1} \geq 0, \ \ x_n + c_{n-1}x_{n-1} + \ldots + c_1 x_1 \geq 0\}.$$

We prove that Ω is positively invariant, i.e., if $x(t_0) \in \Omega$, then $x(t) \in \Omega$, $\forall\, t \geq t_0$.

Note that for $j = 1, \dots, n-1$ and for

$$\begin{aligned} &x_j(\tau) = 0, \quad x_i(\tau) > 0, \quad \forall i \neq j, \ i \leq n-1, \\ &x_n(\tau) + c_{n-1}x_{n-1}(\tau) + \ldots + c_1 x_1(\tau) > 0. \end{aligned}$$

the following inequality holds

$$\dot{x}_j(\tau) > 0. \tag{3.71}$$

Really, for $j = 1, \dots, n-2$ we have

$$\dot{x}_j(\tau) = x_{j+1}(\tau) > 0,$$

for $n = 2$

$$\dot{x}_1(\tau) = x_2(\tau) > -c_1 x_1(\tau) = 0,$$

and for $n > 2$

$$\dot{x}_{n-1}(\tau) = x_n(\tau) > -c_{n-2}x_{n-2}(\tau) - c_1 x_1(\tau) \geq 0.$$

Note also that

$$\begin{aligned} &\left(x_n(\tau) + c_{n-1}x_{n-1}(\tau) + \ldots + c_1 x_1(\tau)\right)^{\bullet} = \\ &= (-a_{n-1} + c_{n-2} + (a_n - c_{n-1})c_{n-1})x_{n-1}(\tau) + \\ &+ \ldots + (-a_2 + c_1 + (a_n - c_{n-1})c_2)x_2(\tau) + \\ &+ (-a_1 + (a_n - c_{n-1})c_1)x_1(\tau). \end{aligned}$$

Then by condition 2) of theorem we obtain inequality.

$$\left(x_n(\tau) + c_{n-1}x_{n-1}(\tau) + \ldots + c_1 x_1(\tau)\right)^{\bullet} > 0 \tag{3.72}$$

for $x_n(\tau) + c_{n-1}x_{n-1}(\tau) + \ldots + c_1 x_1(\tau) = 0$ and $x_j(\tau) > 0$, $j = 1, \dots, n-1$.

From inequalities (3.71) and (3.72) it follows that almost everywhere the boundary of set Ω is a boundary without contact (transversal) with respect to the vector field of system (3.70) and the solutions of system (3.70) intersect almost everywhere the boundary of set Ω. In this case the continuous dependence of solutions of system (3.70) on initial data implies that the set Ω is positively invariant. The positive invariance of Ω yields the lack of asymptotic stability of system (3.70). The proof of theorem is completed.

It is well known [3, 4] another instability condition of system (3.43), namely

$$\mathrm{Tr}\left(A + bK(t)c^*\right) \geq \alpha > 0, \quad \forall t \in \mathbb{R}^1$$

The results obtained are applied to the case that $n = 2$, b, c are vectors, and $K(t)$ is a scalar function.

Consider a transfer function of system (3.43)

$$W(p) = c^*(A - pI)^{-1}b = \frac{\rho p + \gamma}{p^2 + \alpha p + \beta},$$

where p is a complex variable.

Put $\rho \neq 0$. Then without loss of generality we assume that $\rho = 1$. Suppose also that the function $W(p)$ is nondegenerate, i.e., the inequality $\gamma^2 - \alpha\gamma + \beta \neq 0$ is true. In this case system (3.43) may be written as

$$\begin{aligned} \dot{\sigma} &= \eta \\ \dot{\eta} &= -\alpha\eta - \beta\sigma - K(t)(\eta + \gamma\sigma). \end{aligned} \tag{3.73}$$

By the constant $K(t) \equiv K_0$ a stabilization of system (3.73) is possible in the case that $\alpha + K_0 > 0$, $\beta + \gamma K_0 > 0$. For the existence of a number K_0, satisfying these two inequalities, it is necessary and sufficient for either the condition $\gamma > 0$ or the inequalities $\gamma \leq 0$, $\alpha\gamma < \beta$ to be satisfied.

Consider the case that by the constant $K(t) \equiv K_0$ the stabilization is impossible: $\gamma \leq 0$, $\alpha\gamma > \beta$. We apply Theorem 3.11. Condition 1) of Theorem 3.11 is satisfied since $\det bK_0c^* = K_0 \det bc^* = 0$ and $\mathrm{Tr}\, bK_0c^* = -K_0 \neq 0$.

Condition 2) of Theorem 3.11 is satisfied if for some U_0 the polynomial $p^2 + \alpha p + \beta + U_0(p + \gamma)$ has complex zeros. We see that for the existence of such U_0 it is necessary and sufficient that the inequality holds

$$\gamma^2 - \alpha\gamma + \beta > 0 \tag{3.74}$$

Thus, if inequality (3.74) is valid, then there exists a periodic function $K(t)$ such that system (3.73) is asymptotically stable.

Now we obtain the same result by means of Theorem 3.14.

Without loss of generality it can be assumed that $\alpha > 0$. For this purpose it is sufficient to choose an acceptable K_0 in the expression $-(\alpha + K_0)\eta - (\beta + \gamma K_0)\sigma - (K(t) - K_0)(\eta + \gamma\sigma)$ and to make the following change

$\alpha + K_0 \to \alpha$, $\beta + \gamma K_0 \to \beta$, $K(t) - K_0 \to K(t)$. From the inequality $\alpha > 0$ it follows that $\lambda > æ$. Here

$$\frac{c^* b}{\lim\limits_{p \to æ} (æ - p) W(p)} = \frac{æ + \lambda}{æ + \gamma}.$$

Thus all the assumptions of Theorem 3.14 are valid in the case that

$$(\lambda - \gamma)(æ + \gamma) = -\gamma^2 + \alpha\gamma - \beta < 0.$$

This inequality coincides with inequality (3.74). If the inequality

$$\gamma^2 - \alpha\gamma + \beta < 0 \qquad (3.75)$$

is satisfied, then the assumptions of Theorem 3.16 are obviously satisfied too. Thus, we have the following result.

Theorem 3.17 [27]. *If inequality* (3.74) *is valid, then there exists a periodic function* $K(t)$ *such that system* (3.73) *is asymptotically stable.*

If inequality (3.75) *is valid, then there does not exist a function* $K(t)$ *such that system* (3.73) *is asymptotically stable.*

For other class of the stabilizing functions $K(t)$ of the form $K(t) = (k_0 + k_1\omega \cos\omega t)$, $\omega \gg 1$, this result was also obtained in [33] by means of the averaging method.

Consider now several systems of the third order with various transfer functions.

1) $W(p) = \dfrac{1}{p^3 + \alpha p^2 + \beta p + \gamma}$,

where α, β, γ are some numbers.

If $\alpha > 0$, $\beta > 0$, then the stationary stabilization is possible. Let $\alpha > 0$, $\beta \leq 0$. In this case the stationary stabilization is impossible. We now apply Theorem 3.15.

Obviously, for sufficiently large U_0 there are one negative and two complex roots $\lambda_0 \pm i\omega_0$, $\lambda_0 > 0$ of the polynomial $p^3 + \alpha p^2 + \beta p + U_0 + \gamma$.

Take K_1 such that the relation

$$p^3 + \alpha p^2 + \beta p + K_1 + \gamma = (p - æ_1)(p^2 + \alpha_1 p + \beta_1)$$

is valid for $\alpha_1 = \alpha + æ_1$, $\beta_1 = \beta + (\alpha + æ_1)æ_1$. For large $æ_1$ the polynomial $p^2 + \alpha_1 p + \beta_1$ has complex roots with the real part $-(\alpha + æ_1)/2$. Take K_2 such that the relation

$$p^3 + \alpha p^2 + \beta p + K_2 + \gamma = (p + \lambda_2)(p^2 + \alpha_2 p + \beta_2)$$

is satisfied for $\alpha_2 = \alpha - \lambda_2$, $\beta_2 = \beta - (\alpha - \lambda_2)\lambda_2$. Then for large λ_2 the polynomial $p^2 + \alpha_2 p + \beta_2$ has complex roots with the real part $(\lambda_2 - \alpha)/2$.

We obtain $\dim M_1 = \dim L_2 = 1$, $\dim M_2 = \dim L_1 = 2$,

$$\lambda_1 = \frac{\alpha + æ_1}{2}, \quad æ_2 = \frac{\lambda_2 - \alpha}{2},$$

$$\lambda_1\lambda_2 - æ_1 æ_2 = \alpha(\lambda_2 + æ_1) > 0.$$

Thus, all the assumptions of Theorem 3.15 are satisfied.

Since $\operatorname{Tr}(A + bK(t)c^*) = -\alpha$, the asymptotic stability is impossible for $\alpha < 0$.

Thus, we may state the following result.

Theorem 3.18. *For $\alpha > 0$ the system is stabilizable. For $\alpha < 0$ the stabilization is impossible.*

2) $W(p) = \dfrac{p}{p^3 + \alpha p^2 + \beta p + \gamma}$.

For $\alpha > 0$, $\gamma > 0$ the stationary stabilization is possible. Consider the case $\alpha > 0$, $\gamma < 0$ and apply Theorem 3.13 with $K_1 = K_2$, $\lambda_1 = \lambda_2 = \lambda$, $æ_1 = æ_2 = æ$. Take K_1 such that the relation

$$(p - æ)(p^2 + \alpha_1 p + \beta_1) = p^3 + \alpha p^2 + (K_1 + \beta)p + \gamma$$

is satisfied with $\alpha_1 = \alpha + æ$, $\beta_1 = -\gamma/æ$. For small æ the polynomial $p^2 + \alpha_1 p + \beta_1$ has complex zeros with the real part $-(\alpha + æ)/2$. Then $M_1 = M_2$, $L_1 = L_2$, $\dim M_1 = 1$, $\dim L_2 = 2$, $\lambda = (\alpha + æ)/2$. It is clear that for small æ the inequality $\lambda > æ$ holds. Since $\operatorname{Tr}(A + bK(t)c^*) = -\alpha$ for $\alpha < 0$, the asymptotic stability is impossible.

Therefore in this case we can state the following result.

Theorem 3.19. *Let $\alpha \neq 0$, $\gamma \neq 0$. Then for stabilization it is necessary and sufficient that $\alpha > 0$.*

3) $W(p) = \dfrac{p^2}{p^3 + \alpha p^2 + \beta p + \gamma}$.

For $\beta > 0$, $\gamma > 0$ the stationary stabilization is possible. In the case $\beta < 0$, $\gamma < 0$ by Theorem 3.16 the stabilization is impossible.

Consider the case $\beta > 0$, $\gamma < 0$. We apply Theorem 3.9. Put $K_1 = K_2$, $\lambda_1 = \lambda_2 = \lambda$, $æ_1 = æ_2 = æ$.

Take K_1 such that the relation

$$(p - æ)(p^2 + \alpha_1 p + \beta_1) = p^3 + (\alpha + K_1)p^2 + \beta p + \gamma$$

is valid. Here

$$\alpha_1 = \frac{-(\gamma + æ\beta)}{æ^2}, \quad \beta_1 = -\frac{\gamma}{æ},$$

where æ is a small number. Whence it follows that λ can be determined in the following way

$$\lambda = \frac{-(\gamma + æ\beta)}{2æ^2} - \sqrt{\frac{(\gamma + æ\beta)^2}{4æ^4} + \frac{\gamma}{æ}}\,.$$

Obviously, inequality $\lambda > æ$ is satisfied for $b + æ^2 > 0$.

Thus, we can conclude that condition (3.52) is satisfied.

Further we choose U_0 such that the relation

$$(p - \nu)(p^2 + \alpha_2 p + \beta_2) = p^3 + (\alpha + U_0)p^2 + \beta p + \gamma$$

is valid. Here

$$\alpha_2 = \frac{-(\gamma + \nu\beta)}{\nu^2}, \quad \beta_2 = -\frac{\gamma}{\nu},$$

where ν is a sufficiently large number.

Without loss of generality we can assume that

$$A + U_0 bc^* = \begin{pmatrix} \nu & 0 \\ 0 & Q \end{pmatrix}, \quad b = \begin{pmatrix} b_1 \\ b_2 \end{pmatrix}, \quad c = \begin{pmatrix} c_1 \\ c_2 \end{pmatrix}.$$

Here Q is a 2×2-matrix, $b_2 \in \mathbb{R}^2$, $c_2 \in \mathbb{R}^2$. Note that the following relations

$$\frac{c^* b}{c_1 b_1} = \frac{-1}{\lim\limits_{p \to \nu} (\nu - p)W(p)} = 1 + \frac{\alpha_2 \nu + \beta_2}{\nu^2} = 1 - \frac{2\gamma}{\nu^3} - \frac{\beta}{\nu^2} < 1 \qquad (3.76)$$

are valid for $\beta\nu > -2\gamma$.

The matrix Q has complex eigenvalues. Therefore for nonzero vector $u \in M_1$ there exists a number $\tau_1 > 0$ such that the vectors

$$b, \quad \begin{pmatrix} 1 \\ 0 \end{pmatrix}, \quad \left[\exp(A + U_0 bc^*)\tau_1\right]u$$

belong to the same plane.

By Lemma 3.8 inequality (3.76) implies that there exist numbers μ and $\tau(\mu)$ such that

$$\begin{pmatrix} 1 \\ 0 \end{pmatrix}^* \left[\exp(A + (U_0 + \mu)bc^*)\tau(\mu)\right]\left[\exp(A + U_0 bc^*)\tau_1\right]u = 0.$$

From the fact that planes L_2 and $\{x|\, x^* \binom{1}{0} = 0\}$ intersect and a matrix Q has complex eigenvalues it follows that there exists a number $\tau_2 > 0$ such that

$$\big[\exp(A + U_0 bc^*)\tau_2\big]\big[\exp(A + (U_0 + \mu)bc^*)\tau(\mu)\big]\big[\exp(A + U_0 bc^*)\tau_1\big]u \in L_2.$$

This inclusion results in that condition (3.51) holds for the function

$$\begin{aligned}
U(t) &= U_0, & &\forall\, t \in [0, \tau_1),\\
U(t) &= U_0 + \mu, & &\forall\, t \in [\tau_1, \tau_1 + \tau(\mu)),\\
U(t) &= U_0, & &\forall\, t \in [\tau_1 + \tau(\mu), \tau_1 + \tau(\mu) + \tau_2),
\end{aligned}$$

where $\tau = \tau_1 + \tau(\mu) + \tau_2$.

Finally, we may state the following result.

Theorem 3.20. *Let $\beta \neq 0$, $\gamma < 0$. Then for stabilization it is necessary and sufficient that $\beta > 0$.*

We hope that the mathematical tools developed here allow us to obtain in the sequel the other criteria of stabilization.

Chapter 4

Two-dimensional control systems. Phase portraits

In the case that the dimension of a phase space (a state space) equals two, applying some auxiliary tools, we can demonstrate the partition of the phase plane into trajectories of differential equations, which correspond to one or the other of operating regimes of a control system. This allows us to describe qualitatively (and sometimes quantitatively) both the operating regimes of system and its transient processes.

4.1 An autopilot and spacecraft orientation system

Consider a system of angular orientation. A classic example of such a system is a two-positional autopilot.

Let us obtain an equation of a ship rotation around the vertical axis passing through its center of gravity. (We neglect the lateral drift of a ship in the process of rotational motion and assume that a ship moves with a constant velocity.)

$$I\,\frac{d^2\theta}{dt^2}+\alpha\,\frac{d\theta}{dt}=M(\psi). \qquad (4.1)$$

Here $\theta(t)$ is a ship deviation from a given course, I is a ship moment of inertia with respect to its vertical axis, the value $\alpha\,\dot{\theta}(t)$ corresponds to a moment of friction forces, α is a friction factor, $M(\psi)$ is a moment of forces of a rudder, ψ is a rudder blade angular deflection (Fig. 4.1).

Note that if the ship is uncontrollable ($\psi\equiv 0$ and, consequently, $M=0$), then we obtain the trajectories of the corresponding two dimensional system

$$\dot{\theta} = x,$$
$$\dot{x} = -\frac{\alpha}{I}\,x \tag{4.2}$$

in the phase plane $\{x, \theta\}$. In addition, for the solutions $x(t)$, $\theta(t)$ of system (4.2) an identity

$$\left(x(t) + \frac{\alpha}{I}\,\theta(t)\right)^{\bullet} = \dot{x}(t) + \frac{\alpha}{I}\,x(t) \equiv 0$$

is satisfied.

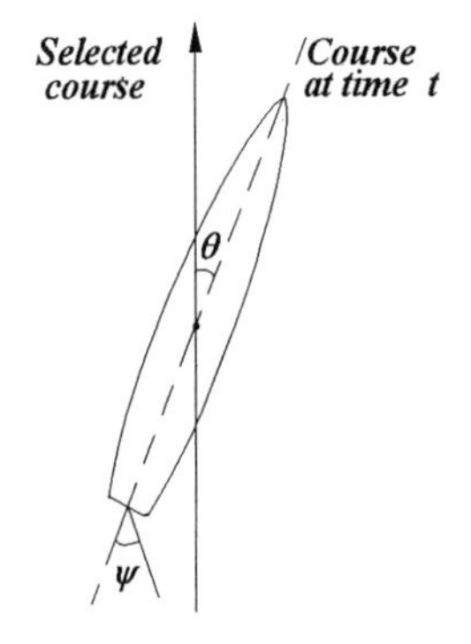

Fig. 4.1

Therefore for any solution of system (4.2) we have

$$x(t) + \frac{\alpha}{I}\,\theta(t) \equiv \text{const}\,.$$

Whence it follows that each straight line

$$x + \frac{\alpha}{I}\,\theta = \gamma$$

consists wholly of three trajectories, one of which is an equilibrium $x = 0$, $\theta = \gamma I/\alpha$ and the two other of trajectories tending to this equilibrium (Fig. 4.2).

Thus, the abscissa axis consists wholly of equilibria.

To each ship movement with initial data

$$\theta(0) = \theta_0, \qquad x(0) = \dot{\theta}(0) = x_0$$

assigns the trajectory with the same initial data. Therefore the rotation of an uncontrolled ship decays with time and as $t \to +\infty$ we have

$$\theta(t) \to \theta_0 + \frac{I}{\alpha} x_0.$$

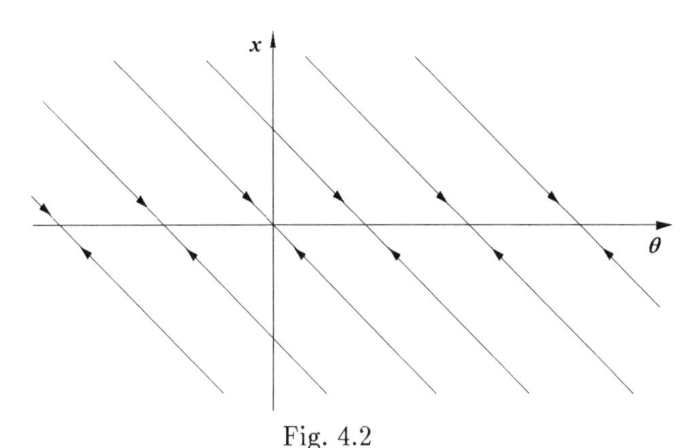

Fig. 4.2

The purpose of control is to ensure conditions such that the following relation

$$\lim_{t \to +\infty} \theta(t) = 0$$

is satisfied in a given region of initial data or we have

$$\lim_{t \to +\infty} \theta(t) = 2k\pi$$

for almost all trajectories. Here the number k depends on the initial data $k = k(\theta_0, x_0)$.

Consider a two-positional autopilot such that the rudder is in two positions only: $\psi = \psi_0$ and $\psi = -\psi_0$. In this case we have moments of forces equal in value but opposite in directions.

At present there are various sensing devices (gyrocompasses), which measure quantities $\theta(t)$ and $\dot{\theta}(t)$, and transmit a signal $\sigma(t) = \theta(t) + b\,\dot{\theta}(t)$ to an actuator, which in the ideal case handles instantly the rudder depending on $\sigma(t)$. Here b is a positive number. Put

$$\begin{aligned}
&\psi(\sigma) = -\psi_0 && \text{for} \quad \sigma \in (0, \pi), \\
&\psi(\sigma) = \psi_0 && \text{for} \quad \sigma \in (-\pi, 0), \\
&\psi(\sigma + 2\pi) = \psi(\sigma), && \forall\, \sigma \in \mathbb{R}^1.
\end{aligned}$$

If $\sigma = 0$ or $\sigma = \pi$, then the actuator is turned off and the rudder turns out to be in any position between $-\psi_0$ and ψ_0, that is, $\psi \in [-\psi_0, \psi_0]$.

We see that the graph of a 2π-periodic function $M(\sigma) = M(\psi(\sigma))$ has the form shown in Fig. 4.3.

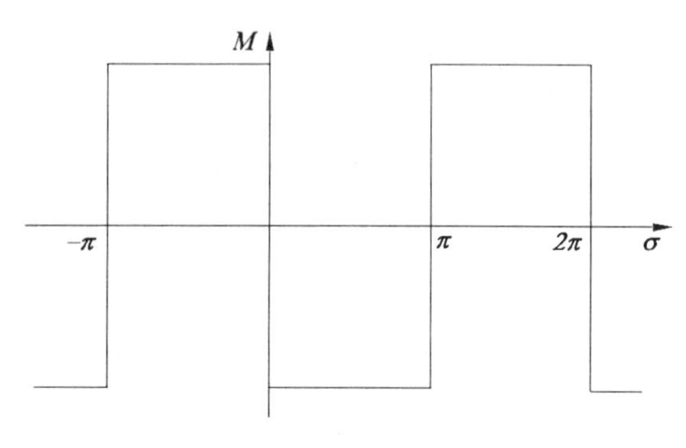

Fig. 4.3

The detailed description of technical realization of such a two-position autopilot can be found in [2].

Note also that a similar problem arises in design of autopilot for an aeroplane. In this case an autopilot must control not one but three angles. The same problem occurs also in the spacecraft orientation systems. However in many cases such more complicated pilot-controlled motions can be regarded as a set of independent plane rotations, in which this case the angle controls is performed by three independent feedbacks. The peculiarity of a spacecraft orientation system is that the forces of viscous resistance ($\alpha = 0$) lack and jet engines can be used as an actuator rather than a rudder. In order to save a fuel the so-called dead zones are introduced, whose sizes are acceptable for the given orientation. In this case the graph of 2π-periodic function $M(\sigma)$ is as follows (Fig. 4.4.)

Here the interval $(-\Delta, \Delta)$ is a dead zone in which the jet engine is put out.

Thus, an autopilot and a spacecraft angular orientation system are described by the following equations

$$I\ddot{\theta} + \alpha\dot{\theta} = M(\sigma), \qquad \sigma = \theta + b\dot{\theta}, \tag{4.3}$$

where numbers α, Δ, b are such that $\alpha \geq 0$, $\Delta \geq 0$, $b > 0$. The graph of $M(\sigma)$ is shown in Fig. 4.4.

Having performed the substitution $\eta = \dot{\theta}$ into system (4.3), we obtain

$$\begin{aligned}\dot{\eta} &= -a\eta + f(\sigma),\\ \dot{\sigma} &= \beta\eta + b\,f(\sigma),\end{aligned} \tag{4.4}$$

where $a = \alpha/I$, $f(\sigma) = M(\sigma)/I$, $\beta = 1 - ab$. Put also $\beta > 0$.

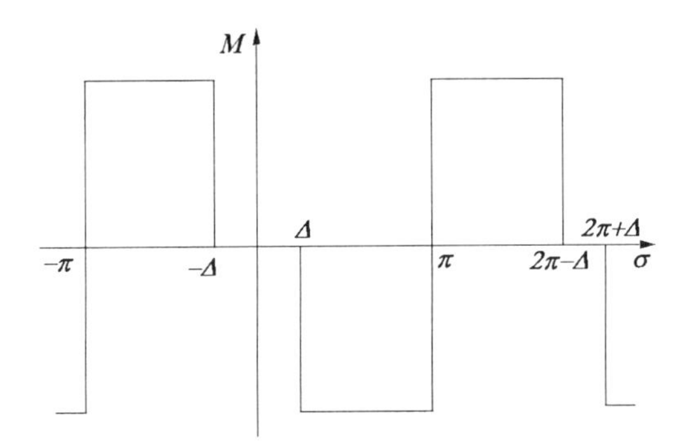

Fig. 4.4

Now we make a remark concerning the definition of solutions of system (4.4) on the lines of discontinuity $\eta \in \mathbb{R}^1$, $\sigma = \Delta + 2\pi k$, $\sigma = -\Delta + 2\pi k$, $\sigma = (2k-1)\pi$. If at time t a solution $\sigma(t)$, $\eta(t)$ is on one of these lines, namely $\sigma(t) = \sigma^*$, then the two different cases are possible:

$$1)\ [\beta\eta(t) + b\,f(\sigma^* - 0)]\,[\beta\eta(t) + b\,f(\sigma^* + 0)] > 0; \tag{4.5}$$

$$2)\ [\beta\eta(t) + b\,f(\sigma^* - 0)]\,[\beta\eta(t) + b\,f(\sigma^* + 0)] \le 0. \tag{4.6}$$

In the first case we assume that at time t a trajectory of system (4.4) "pass" from the semispace $\{\sigma < \sigma^*\}$ into that $\{\sigma > \sigma^*\}$ (or vice versa) (see Fig. 4.5).

Thus, we complete a definition of the solution of system (4.4) on a line of discontinuities by adding the limiting points of the trajectories Γ_1 and Γ_2 for $t - 0$ and $t + 0$ respectively.

In the second case the velocity vectors S_1 and S_2 of trajectories Γ_1 and Γ_2 for $t - 0$ and for $t + 0$ respectively are obtained in the following way (see Fig. 4.6).

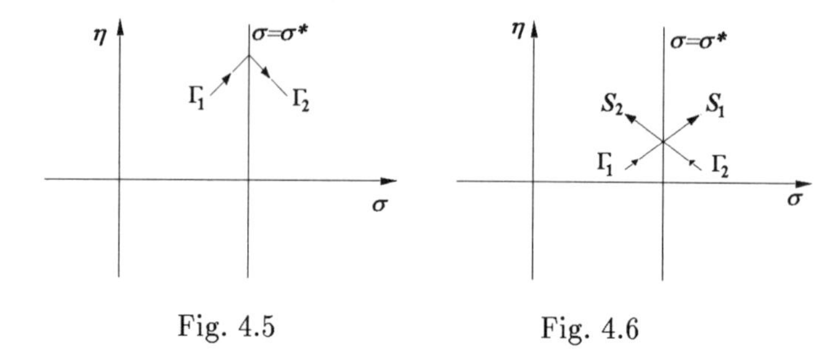

Fig. 4.5 Fig. 4.6

We see that the trajectory Γ_1 is not extensible by continuity into the semispace $\{\sigma > \sigma^*\}$ and the trajectory Γ_2 into that $\{\sigma < \sigma^*\}$. In this case the extended trajectory is also a continuation of trajectories Γ_1 and Γ_2 (here a solution of the Cauchy problem is not unique !). The solution slides along the line of discontinuity $\sigma = \sigma^*$ until inequality (4.6) is violated. The velocity vector S_3 of sliding solution is defined from Fig. 4.7.

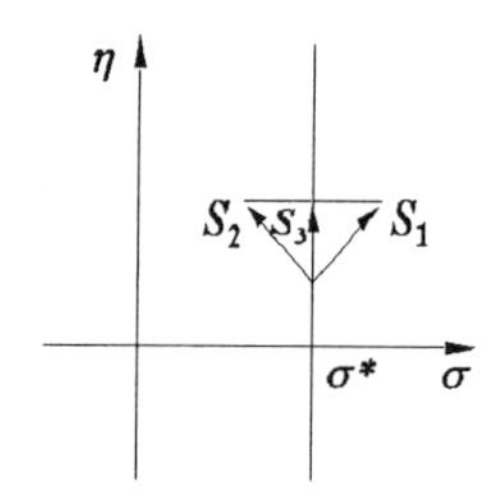

Fig. 4.7

Thus, the end of vector S_3 is a crosspoint of the line $\sigma = \sigma^*$ and a segment, connecting the ends of vectors S_1 and S_2. The vector field so defined determines uniquely the trajectories, placed on the surfaces of discontinuity on the right-hand sides of differential equations. Such solutions are called sliding solutions. This method for supplementing a definition of solutions was suggested by A.F. Filippov and at present it is widely used in the control theory.

For further details, concerning the theory of differential equations and differential inclusions with discontinuous right-hand sides, we refer an interested reader to the books [13, 17].

Note that in the case, considered above, for a sliding solution we obtain $\dot{\sigma}(t) = 0$. Besides, for any solution $\eta(t)$, $\sigma(t)$ the number of entries into and exits from a sliding regime is no more than countable: $t = t_k$, $k = 1, 2, \ldots$.

We note also that since for a sliding solution we have $\dot{\sigma}(t) = 0$, the second equation of system (4.4) implies that on a discontinuity $\{\eta \in \mathbb{R}^1, \ \sigma = \sigma^*\}$ in place of any value of $f(\sigma^*)$ the following relation

$$f(t) = -\frac{\beta}{b}\,\eta(t) \tag{4.7}$$

must be taken. Using the first equation of (4.4), we obtain very simple equations of a sliding regime

$$\sigma(t) = \sigma^*, \qquad \dot{\eta} = -\left(a + \frac{\beta}{b}\right)\eta. \tag{4.8}$$

Note again that the nonlinearity of f is determined in a sliding regime by formula (4.7), which depends on initial data. Such a supplementing of definition is due to the rule of constructing the vector field on a line of discontinuity (see Fig. 4.7). In this case it is important that for all sliding regimes inequality (4.6) is satisfied. From (4.7) it follows that the values of $f(t)$ always belong to the segment $[f(\sigma^* - 0),\ f(\sigma^* + 0)]$ for $f(\sigma^* - 0) < f(\sigma^* + 0)$ and to the segment $[f(\sigma^* + 0),\ f(\sigma^* - 0)]$ for $f(\sigma^* - 0) > f(\sigma^* + 0)$.

Consider a function

$$V(\eta, \sigma) = \eta^2 + q\int\limits_0^\sigma f(\sigma)\,d\sigma, \tag{4.9}$$

where

$$q = -\frac{2(1 + ab)}{(1 - ab)^2}\,.$$

We also consider the derivative of function V with respect to system (4.4), i.e., an expression

$$\begin{aligned}\frac{d}{dt}V\big(\eta(t), \sigma(t)\big) = 2\eta(t)\big(-a\eta(t) + f(\sigma(t))\big) +\\ + q\,f(\sigma(t))\,\big(\beta\eta(t) + b\,f(\sigma(t))\big)\end{aligned} \tag{4.10}$$

for values of t such that $\sigma(t) \neq \sigma^*$. Here σ^* are the points of discontinuity of the function $f(\sigma)$.

Since for a sliding regime $\sigma(t) = \sigma^*$, the following relation holds

$$\int_0^{\sigma(t)} f(\sigma)\, d\sigma = \int_0^{\sigma^*} f(\sigma)\, d\sigma$$

and the function f is completed by formula (4.7), we have

$$\frac{d}{dt} V(\eta(t), \sigma(t)) = -2\left(a + \frac{\beta}{b}\right) \eta(t)^2. \tag{4.11}$$

By (4.10)

$$\frac{d}{dt} V(\eta(t), \sigma(t)) = -2\left[a\eta(t)^2 + \frac{2ab}{(1-ab)} f(\sigma(t))\,\eta(t) + \right.$$

$$\left. + \frac{(1+ab)b}{(1-ab)^2} f(\sigma(t))^2\right] = -2a\left(\eta(t) + \frac{b}{(1+ab)} f(\sigma(t))\right)^2 - \tag{4.12}$$

$$-\frac{2b}{(1-ab)^2} f(\sigma(t))^2.$$

The function V with properties (4.11) and (4.12), constructed above, is the Lyapunov function (see Chapter 1). It is nonnegative for all η and σ. A derivative of function V with respect to system (4.4) is nonpositive and satisfies relations (4.11) and (4.12) everywhere except for the points t_k. Using this function, we show that any solution of system (4.4) tends to a certain equilibrium.

For proving this fact we need in the following statement.

Proposition 4.1. *If $\eta(t)$, $\sigma(t)$ is a solution of system (4.4), then $\eta(t)$, $\sigma(t) + 2k\pi$ is also a solution of system (4.4).*

The proof is in substituting the solution $\eta(t)$ $\sigma(t) + 2k\pi$ into system (4.4), taking into account a 2π-periodicity of function $f(\sigma)$.

It follows from Proposition 4.1 that if we have performed a shift of the band $\{\eta \in \mathbb{R}^1,\ \sigma \in [-\pi, \pi]\}$ along the abscissa axis by $2\pi k$, then the parts of trajectories, placed in this band, coincide with the corresponding parts of trajectories placed in the band $\{\eta \in \mathbb{R}^1,\ \sigma \in [(2k-1)\pi, (2k+1)\pi]\}$. We use this fact for a qualitative description of behavior of trajectories in the phase plane.

Consider now the case of autopilot and sliding regimes in the phase plane (Fig. 4.8). These processes are described by equations (4.8). Using equations (4.7) and taking into account that $f(t) \in [f(\sigma^* - 0),\ f(\sigma^* + 0)]$, we have

$$|\eta(t)| \leq d,$$

where $d = b\,|f(\sigma^* - 0)|\,/\,|\beta|$.

The analysis of the vector field makes it possible to describe the qualitative behavior of trajectories in small neighborhoods of sliding regimes (Fig. 4.9).

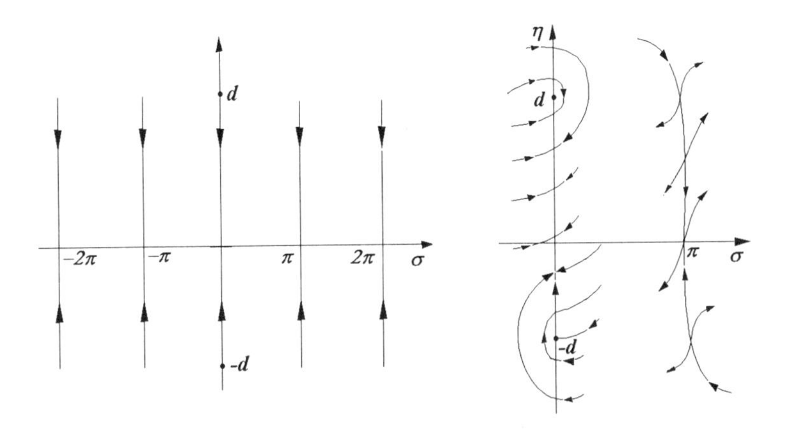

Fig. 4.8 Fig. 4.9

We see from this figure that only two trajectories may enter the sliding regime on the line of discontinuity $\{\eta \in \mathbb{R}^1,\ \sigma = \pi\}$. Each point of the segment $\{\sigma = \pi,\ \eta \in [-d, d]\}$ is a branch point of solutions. From this point the three trajectories are starting only. One of them enters into the right half-plane, the second into the left one and the third trajectory remains in a sliding regime. In other words, the sliding regime consists of three trajectories, namely of an equilibrium $\sigma = \pi$, $\eta = 0$ and of two trajectories, which tend to this equilibrium as $t \to \infty$. Each point on these trajectories is a branch point.

Note that the trajectory $\sigma(t), \eta(t)$, entering the sliding regime $\{\eta \in [-d, d], \sigma = \pi\}$ in time $t = t_0$, has the following property. On the interval $(-\infty, t_0)$ the function $|\sigma(t) - \pi|$ is the steadily decreasing one and $|\eta(t)| > d$.

This fact results from identity (4.12) and from the following relation

$$q\int_{\pi}^{\sigma} f(\sigma)\,d\sigma \leq 0, \quad \forall\, \sigma \in (-\infty, +\infty)$$

(see Fig. 4.10).

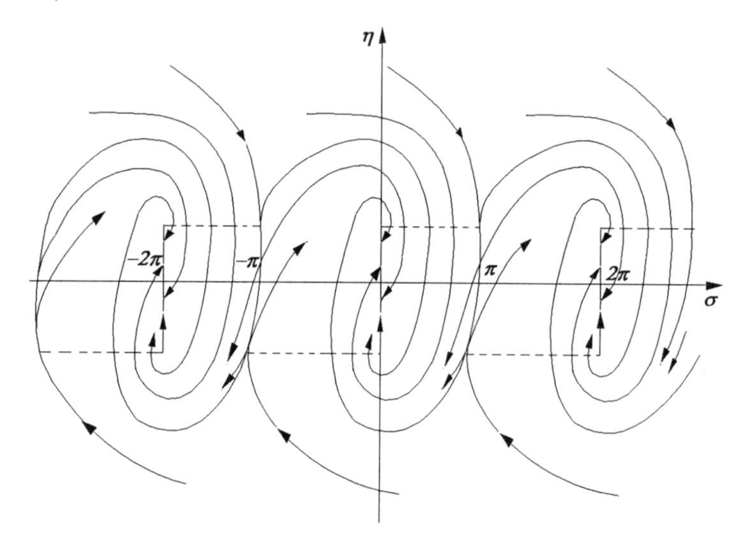

Fig. 4.10

The segment $\{\eta \in [-d, d],\ \sigma = 0\}$ of a sliding process is locally stable: in some ε-neighborhood of this segment in a finite time all the trajectories reach this segment and then remain there, tending to zero as $t \to +\infty$.

Note that a property to reach equilibrium in a finite time is a property of differential equations with discontinuous right-hand sides only. For the systems with smooth right-hand sides it is impossible. In this case the two trajectories reach a zero equilibrium in a finite time.

Suppose now that the trajectory considered does not tend to equilibrium. From the previous reasoning it is clear that except for a countable set of crosspoints t_k of the lines of discontinuity $\{\eta \in \mathbb{R}^1,\ \sigma = \sigma^*\}$ on this trajectory the following inequality holds

$$\big|f(\sigma(t))\big| \geq l,$$

where l is a positive number (see Fig. 4.3). Identity (4.12) implies that the following relation

$$\lim_{t \to +\infty} V\big(\eta(t), \sigma(t)\big) = -\infty$$

is valid. The last relation contradicts the boundedness from below of the function $V(\eta, \sigma)$. The contradiction obtained makes it possible to prove that any solution of system (4.4) tends to equilibrium as $t \to +\infty$.

Hence the following qualitative partition of the phase plane into the trajectories of system (4.4) (Fig. 4.10) can be obtained.

In the control theory this qualitative picture of phase space, filled with trajectories, is called a phase portrait of system.

We now discuss just in the same way as before the spacecraft orientation system.

In this case the main distinction from the autopilot is in the presence of a dead zone and in the absence of a friction ($a = 0$). This fact implies the following changes in Fig. 4.8 and 4.9 (Fig. 4.11 and 4.12).

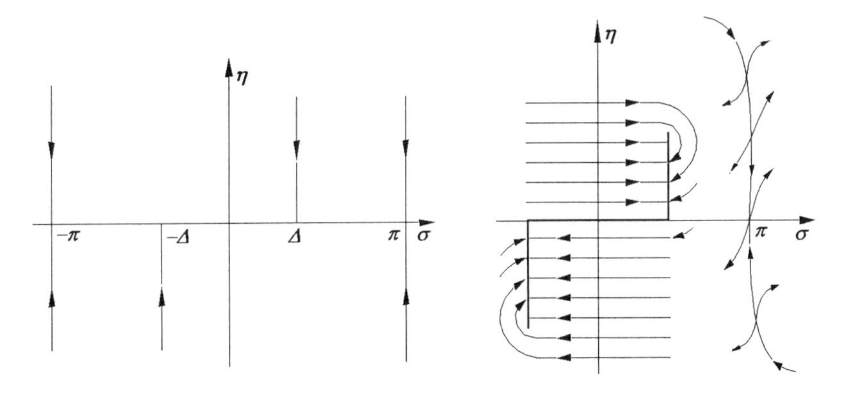

Fig. 4.11 Fig. 4.12

From the configuration of trajectories in Fig. 4.12 we can conclude that the trajectory that does not tend to equilibrium as $t \to +\infty$, as in the case of autopilot, satisfies relation (4.12) everywhere except for a countable set of crosspoints t_k of the lines of discontinuity. Whence it follows that

$$V\big(\eta(t), \sigma(t)\big) \le V\big(\eta(0), \sigma(0)\big).$$

Therefore there exists a number Γ depending on initial data $\eta(0)$, $\sigma(0)$ and such that the following estimate holds

$$|\eta(t)| \leq \Gamma, \qquad \forall t \geq 0. \tag{4.13}$$

We see that for the trajectory considered there exists an infinite sequence $t_j \to +\infty$ such that t_{2i} is an entry time of the trajectory $\eta(t)$, $\sigma(t)$ into the set

$$\Omega = \bigcup_k \{\eta \in \mathbb{R}^1, \quad \sigma \in [-\Delta + 2\pi k, \ \Delta + 2\pi k]\}$$

and t_{2i-1} is an exit time of the trajectory from the set Ω.

Let us introduce the following notation $L = \max_\sigma |f(\sigma)|$. From equations (4.4) it follows that one of the inequalities

$$t_{2i} - t_{2i-1} \geq \frac{d}{L}, \tag{4.14}$$

$$t_{2i} - t_{2i-1} \geq \frac{2(\pi - \Delta)}{\Gamma + bL} \tag{4.15}$$

is true.

Really, we have either $\sigma(t_{2i}) = \sigma(t_{2i-1})$ or the following relation

$$|\sigma(t_{2i}) - \sigma(t_{2i-1})| = 2(\pi - \Delta). \tag{4.16}$$

In the first case we obtain

$$|\eta(t_{2i}) - \eta(t_{2i-1})| > d \tag{4.17}$$

(see Fig. 4.12).

Therefore from the fact that outside $\Omega \cup \{\sigma = (2k-1)\pi\}$ we have $|f(\sigma)| = L$ and from the first equation of system (4.4) (recall that here $\alpha = 0$) it follows that

$$|\eta(t_{2i}) - \eta(t_{2i-1})| = L(t_{2i} - t_{2i-1}).$$

Then from (4.17) estimate (4.14) follows.

In the second case from estimate (4.13) and the second equation of system (4.4) (recall that here $\beta = 1$) we obtain an inequality

$$|\dot{\sigma}(t)| \leq \Gamma + bL, \quad \forall\, t \in (t_{2i-1}, t_{2i}).$$

Relation (4.16) yields estimate (4.15).

Estimates (4.14), (4.15), and (4.12) imply the validity of one of inequalities

$$V\big(\eta(t_{2i}),\sigma(t_{2i})\big) \le V\big(\eta(t_{2i-1}),\sigma(t_{2i-1})\big) - 2bLd,$$
$$V\big(\eta(t_{2i}),\sigma(t_{2i})\big) \le V\big(\eta(t_{2i-1}),\sigma(t_{2i-1})\big) - 2bL^2\,\frac{2(\pi-\Delta)}{\Gamma+bL},$$

Hence

$$\lim_{i\to+\infty} V\big(\eta(t_{2i}),\sigma(t_{2i})\big) = -\infty.$$

This limit contradicts the boundedness from below of the function $V(\eta,\sigma)$. The contradiction makes it possible to prove that any trajectory of system (4.4) tends to equilibrium as $t\to+\infty$. (Note that a more universal method to use the Lyapunov functions for the analysis of stability of discontinuous systems can be found in [17]. However this generalization demands the developed theory of differential inclusions.)

From the fact that any solution of the system tends to equilibrium as $t\to+\infty$ and from the previous analysis of neighborhoods of sliding regimes we may obtain a qualitative partition of the phase plane into trajectories (see Fig. 4.13).

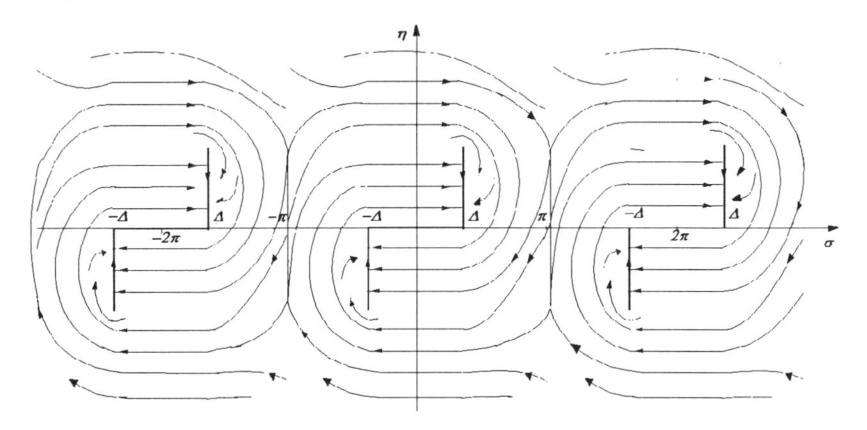

Fig. 4.13

Finally, note that the equilibria $\eta=0$, $\sigma=2\pi k$ of system (4.4) for autopilot correspond to the same given course. All other processes tend to this course asymptotically as $t\to+\infty$ except for special processes, tending to unstable equilibrium $\eta = 0, \sigma = (2k+1)\pi$. However by virtue of instability this process is physically nonrealizable. The similar remarks can be made for the spacecraft orientation control. A systematic presentation of the spacecraft orientation control theory can be found in [38].

It should be noted that the failure of spacecraft orientation system may cause a spacecraft rotation. As it can be seen from the phase portrait (Fig. 4.13) the initiation of the orientation control system may damp this rotation and orient the spacecraft in the proper direction. Such a situation arose at the space station "Mir" in 1987. A breakdown of a control system implies a complicated rotational motion of station. It had taken a certain time to recover the orientation system. After its initiation the orientation control system has damped the rotation of station.

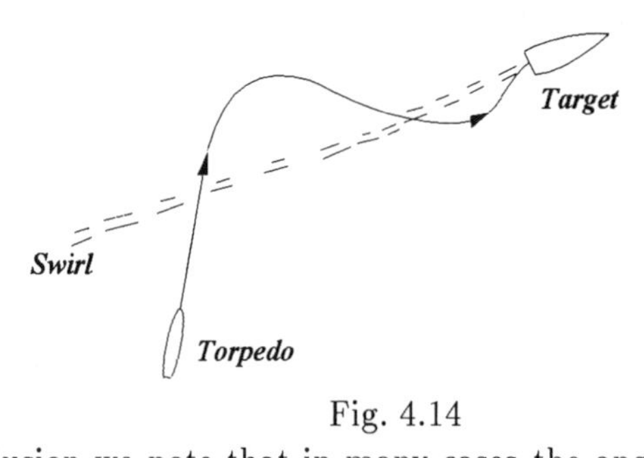

Fig. 4.14

In conclusion we note that in many cases the angular orientation control system must realize not constant direction but a very complicated maneuvering, which is usually called a programming motion. Consider, for example, a torpedoing of a moving ship. The torpedo is directed to not target but to a swirl. This trace can be detected by torpedo devices when the torpedo crosses it. Then the torpedo control system must perform the program motion of a torpedo roll-out and its second entry into the swirl (Fig. 4.14). When crossing a swirl the torpedo makes the second roll-out and so on until the torpedo will reach the target.

4.2 A synchronous electric machine control and phase locked loops

1. Synchronous machine.

Synchronous electric machines are widely spread as current generators. They are often used as electromotors. For example, as a motors for rolling mills.

We discuss here a synchronous electromotor whose main elements are a stationary stator and rotating rotor (Fig. 4.15).

In the figure the rotor slots are shown in which a rotor winding, the so-called direct current excitation winding, is placed. The winding is connected via brushes with a direct current source. On the stator there are also windings for alternating current, which creates an alternating magnetic field. These windings are such that when passing an alternating current the magnetic-field vector is constant by absolute value and rotates with a constant angular velocity (Fig. 4.16).

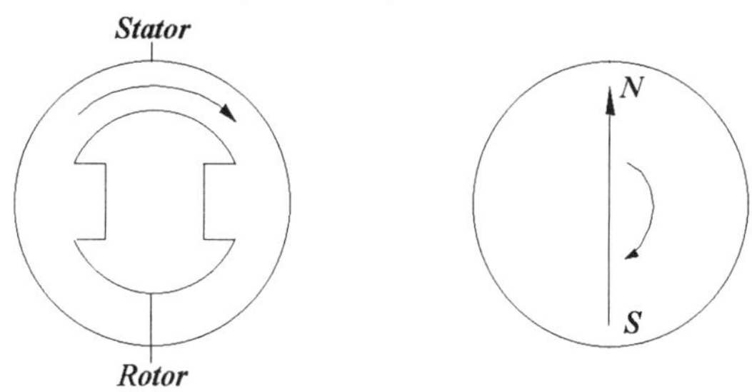

Fig. 4.15 Fig. 4.16

It is clear that this velocity coincides with the frequency of alternating current passing through the stator windings.

Each loop of an excitation winding can be regarded as a loop with the direct current, placed in a magnetic field (Fig. 4.17).

We consider the motion of the loop in the rotating coordinate system. In this coordinate system we have a loop with the direct current i, placed in a constant magnetic field. The force F acts on this loop (Fig. 4.18).

It is well known that $F = \beta i B$, where B is a magnetic intensity, β is a factor of proportionality. Whence it follows that the value $\theta(t)$ satisfies the following equation

$$I\ddot{\theta} = -\beta\, i\, n\, l\, B\, \sin\theta + M. \tag{4.18}$$

Here I is a moment of inertia, $2l$ is a distance between parallel parts of the loop, n is the number of loops, and M is a drag torque. Often, M is called an electromotor load.

Let us consider equation (4.18) in the form

$$\ddot{\theta} + b\sin\theta = \gamma. \tag{4.19}$$

Recall that for $\gamma = 0$ we already considered this equation: it is an equation of a pendulum motion.

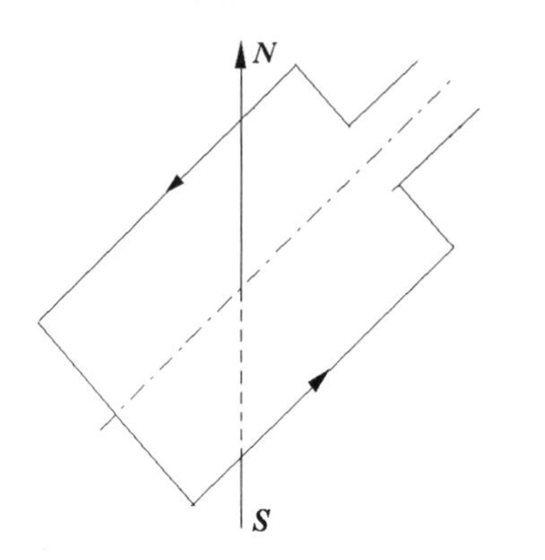

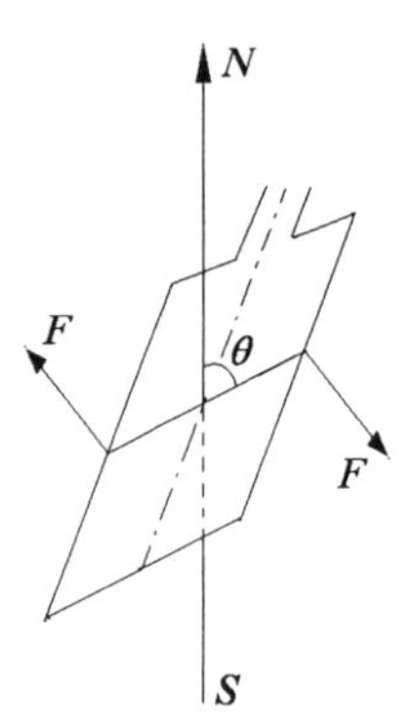

Fig. 4.17 Fig. 4.18

Equation (4.19) is equivalent to a system

$$\begin{aligned} \dot{\theta} &= \eta, \\ \dot{\eta} &= -b\sin\theta + \gamma, \end{aligned} \tag{4.20}$$

which has the first integral

$$V(\eta, \theta) = \frac{1}{2}\eta^2 - b\cos\theta - \gamma\theta.$$

Since

$$\begin{aligned} &\dot{V}(\eta(t), \theta(t)) = \eta(t)\dot{\eta}(t) + (b\sin\theta - \gamma)\dot{\theta}(t) = \\ &= \eta(t)[-b\sin\theta + \gamma + b\sin\theta - \gamma] = 0, \end{aligned}$$

the trajectories of system (4.20) are wholly placed on the level curves of the function $V(\eta, \theta) = C$ (Fig. 4.19). We consider here the most substantial case $b > \gamma$. For $b < \gamma$ the synchronous machine load is so large that in this case it is impossible to ensure the synchronous processes.

Let us now determine what rotor motions correspond to one or other trajectory. The closed trajectories correspond to the motions that can be divided into two components, namely the rotations with a constant angular speed are equal to a magnetic vector speed and periodic swings of rotor with respect to this uniform rotation.

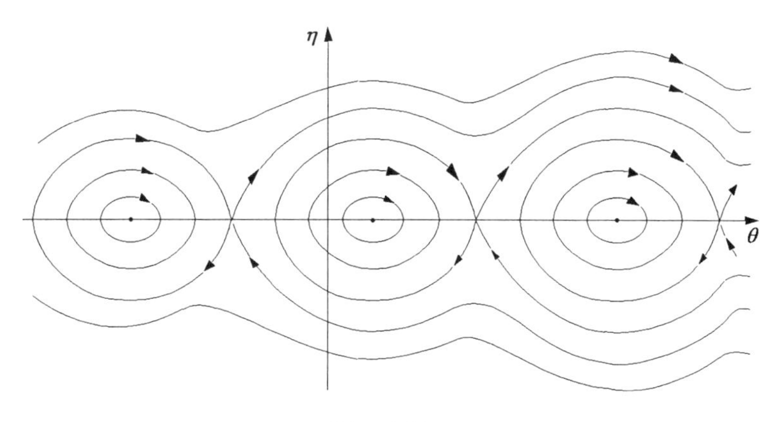

Fig. 4.19

The nonclosed trajectories correspond to the asynchronous motions of rotor. In course of time its angular speed becomes less than a magnetic field speed. Such processes are forbidden for a synchronous machine.

Figure 4.19 show that these two types of trajectories are separated by peculiar trajectories, which are also called homoclinic trajectories. As $t \to +\infty$ and $t \to -\infty$ these trajectories tend to the same stationary saddle point. Sometimes, to underline their separating role, they are also called separatrices.

The objective of a synchronous machine control is to synchronize rotation of rotor and a rotation of magnetic field. For this purpose it was necessary to create the control devices, which could depress (damp) the above-mentioned swinging of rotor with respect to a rotating magnetic field and could expand regions, filled with the bounded solutions, (see Fig. 4.19).

The first such devices were invented in the early 20th century. They were very simple by construction and consisted of two additional short-circuited windings placed on the rotor. A motion of rotor with respect to the magnetic field induced a current in the windings, which, in turn, created forces acting on the rotor and damping its swings.

The inventing of such control devices, which is brilliant by simplicity and technology, may be compared with the invention of the Watt regulator.

The complete mathematical model of the process of interaction of damping windings (damper bar) with the machine is described by a system of differential equations of higher order. We refer an interested reader to the book [48]. There is also engineering argumentation, which we omit here, for a support of the assertion that a moment of damping forces, caused by additional windings, equals $-\alpha\dot{\theta}$, where α is a constant parameter. In this case in place of equation (4.19) we obtain

$$\ddot{\theta} + \alpha\dot{\theta} + b\sin\theta = \gamma. \tag{4.21}$$

Without loss of generality it can be assumed that $b = 1$. The equation of this type can be obtained from equation (4.21), using a change of variable $\tau = t\sqrt{1/b}$. A qualitative study of equation (4.21) by the remarkable Italian mathematician Franchesko Tricomi in 1933 was primarily due to the problem of investigating a synchronous machine operating with the moment of damping force.

We discuss the main results of Tricomi. Consider the following system equivalent to (4.21)

$$\begin{aligned} \dot{\theta} &= \eta, \\ \dot{\eta} &= -\alpha\eta - \varphi(\theta), \end{aligned} \tag{4.22}$$

where $\varphi(\theta) = \sin\theta - \gamma$.

Theorem 4.1. *Any bounded for $t \geq 0$ solution of system (4.22) tends to some equilibrium as $t \to +\infty$.*

Proof. Consider a function

$$V(\eta, \theta) = \frac{1}{2}\eta^2 + \int_0^\theta \varphi(\theta)\, d\theta. \tag{4.23}$$

On the solutions $\eta(t)$, $\theta(t)$ of system (4.22) we have

$$\dot{V}(\eta(t), \theta(t)) = -\alpha\eta(t)^2. \tag{4.24}$$

Whence it follows that the function $V(\eta(t), \theta(t))$ is a monotonically decreasing function of t. By virtue of boundedness of $\eta(t)$, $\theta(t)$ for $t \geq 0$ the function $V(\eta(t), \theta(t))$ is also bounded. This fact implies the existence of a finite limit

$$\lim_{t\to+\infty} V(\eta(t), \theta(t)) = æ. \tag{4.25}$$

Integrating both sides of equation (4.24) from 0 to t, we obtain

$$\alpha \int_0^t \eta(\tau)^2 d\tau = V(\eta(0), \theta(0)) - V(\eta(t), \theta(t)).$$

Hence

$$\int_0^{+\infty} \eta(t)^2 dt < \infty. \tag{4.26}$$

In addition, we have

$$[\eta(t)^2]^\bullet = 2\eta(t)\dot{\eta}(t) = -2\alpha\eta(t)^2 - 2\eta(t)\varphi(\theta(t)).$$

From the boundedness of the solution $\eta(t)$, $\theta(t)$ it follows that there exists a number C such that

$$\left|[\eta(t)^2]^\bullet\right| \leq C, \qquad \forall\, t \geq 0. \tag{4.27}$$

By (4.26), (4.27), and Lemma 1.3 (Chapter 1) we obtain

$$\lim_{t \to +\infty} \eta(t) = 0. \tag{4.28}$$

Therefore, from relation (4.25) and expressions for the functions $V(\eta, \theta)$ and $\varphi(\theta)$ it follows that a number β can be found such that

$$\lim_{t \to +\infty} (\cos\theta(t) + \gamma\theta(t)) = \beta.$$

Then for a certain number θ_0 we have

$$\lim_{t \to +\infty} \theta(t) = \theta_0. \tag{4.29}$$

We see that the point $\eta = 0$, $\theta = \theta_0$ is an equilibrium of system (4.22). The statement of Theorem 4.1 follows from relations (4.28) and (4.29).

Theorem 4.1 implies that for $1 < \gamma$, i.e., in the absence of equilibrium, all the solutions are unbounded (Fig. 4.20).

Now we consider the case $1 > \gamma$.

Note that since function (4.23) satisfies inequality (4.24), the level curves of the function $V(\eta, \theta)$ for $\eta \neq 0$ are the curves without contact (transversal) for trajectories of system (4.22). In other words, these trajectories intersekt level curves from outside toward the interior (Fig. 4.21).

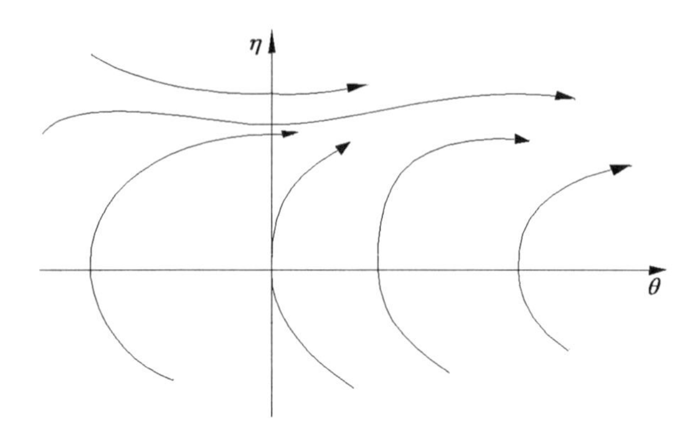

Fig. 4.20

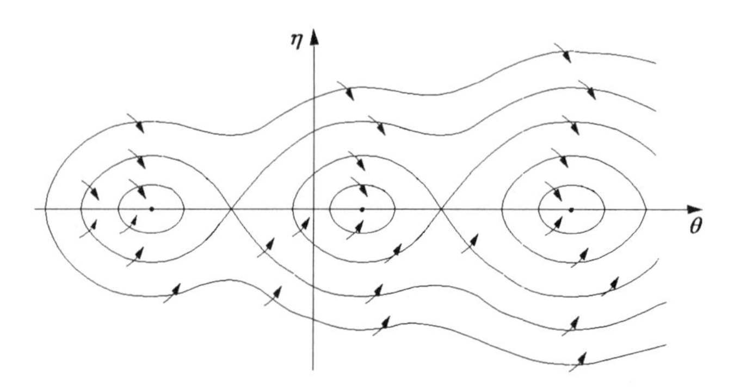

Fig. 4.21

Let us denote by θ_0 a zero of function $\varphi(\theta)$ such that $\theta_0 \in [0, 2\pi]$, $\varphi'(\theta_0) > 0$ and by θ_1 a zero of function $\varphi(\theta)$ such that $\theta_1 \in [0, 2\pi]$, $\varphi'(\theta_1) < 0$. Denote also by $\widetilde{\eta}(t)$, $\widetilde{\theta}(t)$ a solution of system (4.22) such that

$$\lim_{t \to +\infty} \widetilde{\eta}(t) = 0, \qquad \lim_{t \to +\infty} \widetilde{\theta}(t) = \theta_1$$

and $\widetilde{\eta}(t) < 0$ for sufficiently large t.

Let $\widetilde{\widetilde{\eta}}(t)$, $\widetilde{\widetilde{\theta}}(t)$ be a solution of system (4.22) such that

$$\lim_{t \to +\infty} \widetilde{\widetilde{\eta}}(t) = 0, \qquad \lim_{t \to +\infty} \widetilde{\widetilde{\theta}}(t) = \theta_1$$

and $\tilde{\tilde{\eta}}(t) > 0$ for sufficiently large t (Fig. 4.22).

From the fact that a level curve of the function $V(\eta,\theta)$, passing trough the point $\eta = 0$, $\theta = \theta_1$, is the level curve without contact it follows that a trajectory $\tilde{\eta}(t)$, $\tilde{\theta}(t)$ for all $t \in \mathbb{R}^1$ lays below the curve $\{V(\eta,\theta) = V(0,\theta_1),\ \eta < 0\}$ (Fig. 4.23). Since from relations $V(\eta,\theta) = V(0,\theta_1)$ and $\theta \to +\infty$ it follows that $\eta \to -\infty$, we can state the following lemma.

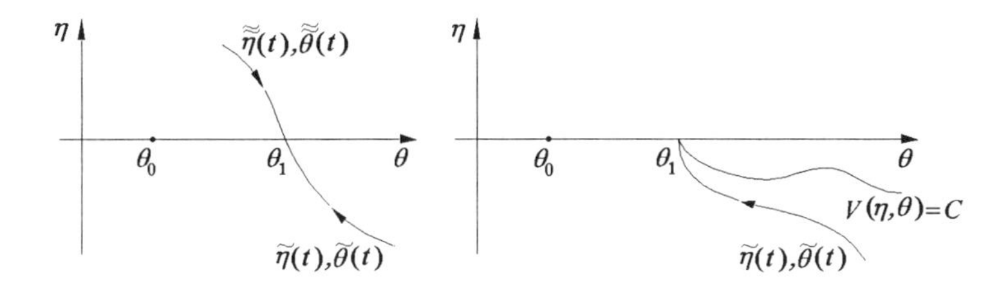

Fig. 4.22 Fig. 4.23

Lemma 4.1. *The following relations*

$$\lim_{t\to-\infty} \tilde{\eta}(t) = -\infty, \qquad \lim_{t\to-\infty} \tilde{\theta}(t) = +\infty \tag{4.30}$$

are valid.

Obviously, there are three possibilities for the trajectory $\tilde{\tilde{\eta}}(t)$, $\tilde{\tilde{\theta}}(t)$:

1) a number τ exists such that $\tilde{\tilde{\eta}}(\tau) = 0$, $\tilde{\tilde{\theta}}(\tau) \in (\theta_1 - 2\pi, \theta_0)$, $\tilde{\tilde{\eta}}(t) > 0$, $\forall\, t \in (\tau, +\infty)$ (Fig. 4.24),

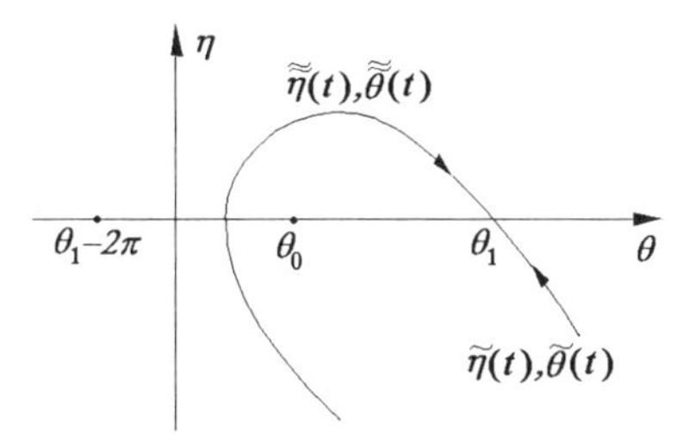

Fig. 4.24

2) for all t we have $\tilde{\tilde{\eta}}(t) > 0$, $\lim_{t\to-\infty} \tilde{\tilde{\eta}}(t) = 0$, and $\lim_{t\to-\infty} \tilde{\tilde{\theta}}(t) = \theta_1 - 2\pi$, (Fig. 4.25),

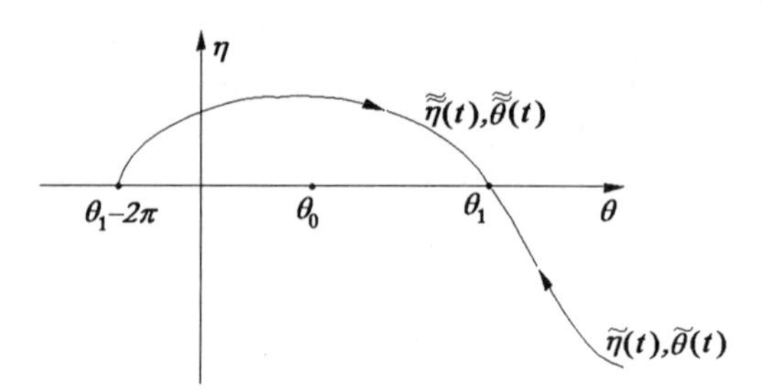

Fig. 4.25

3) for all t we have $\tilde{\tilde{\eta}}(t) > 0$ and $\lim_{t\to-\infty} \tilde{\tilde{\theta}}(t) = -\infty$ (Fig. 4.26).

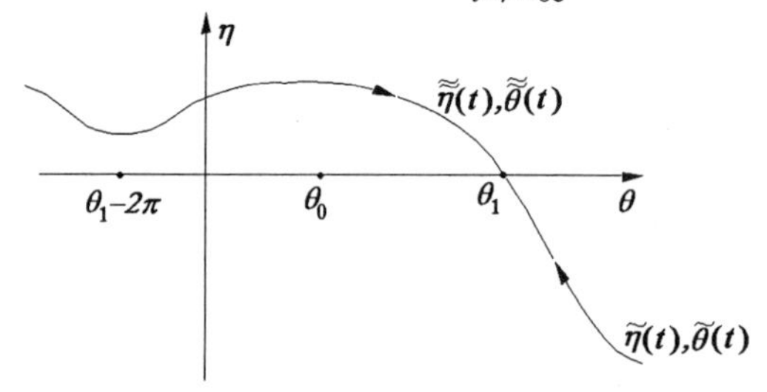

Fig. 4.26

F. Tricomi has shown that in case 3) the following relation holds

$$\lim_{t\to-\infty} \tilde{\tilde{\eta}}(t) = +\infty. \tag{4.31}$$

It is a subtle theorem, the proof of which is omitted and can be found, for example, in [7]. In the works of scientists of the A.A. Andronov school the two-dimensional systems have recently been indicated such that in case 3) relation (4.31) is not valid [19].

From Theorem 4.1 and the properties of separatrices $\tilde{\eta}(t)$, $\tilde{\theta}(t)$ and $\tilde{\tilde{\eta}}(t)$, $\tilde{\tilde{\theta}}(t)$ of the saddle point $\eta = 0$, $\theta = \theta_1$, we can conclude that for $1 > \gamma$ three types of partition of the phase space into trajectories are possible:

1) The separatrices $\tilde{\eta}(t)$, $\tilde{\theta}(t)$ and $\tilde{\tilde{\eta}}(t)$, $\tilde{\tilde{\theta}}(t)$ are boundaries of a domain of attraction for a locally stable equilibrium $\eta = 0$, $\theta = \theta_0$ (Fig. 4.27). When shifting along θ by a value $2k\pi$ these domains turn out to be domains of attraction for stationary solutions $\eta = 0$, $\theta = \theta_0 + 2k\pi$.

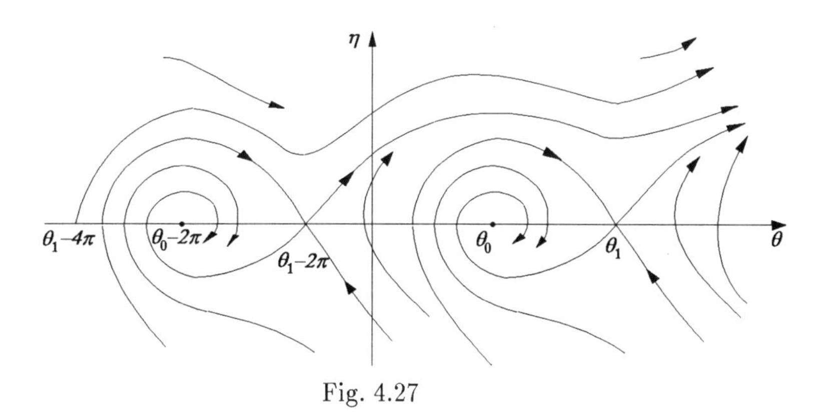

Fig. 4.27

The trajectories outside these domains tend to infinity as $t \to +\infty$.

2) A separatrix $\tilde{\tilde{\eta}}(t)$, $\tilde{\tilde{\theta}}(t)$ is heteroclinic, i.e.,

$$\lim_{t\to-\infty} \tilde{\tilde{\eta}}(t) = \lim_{t\to+\infty} \tilde{\tilde{\eta}}(t) = 0, \quad \lim_{t\to-\infty} \tilde{\tilde{\theta}}(t) = \theta_1 - 2\pi, \quad \lim_{t\to+\infty} \tilde{\tilde{\theta}}(t) = \theta_1.$$

In this case the domains of attraction of stable equilibria are also bounded by separatrices but in the half-plane $\{\eta \leq 0\}$ there do not exist unstable "corridors" (Fig. 4.28).

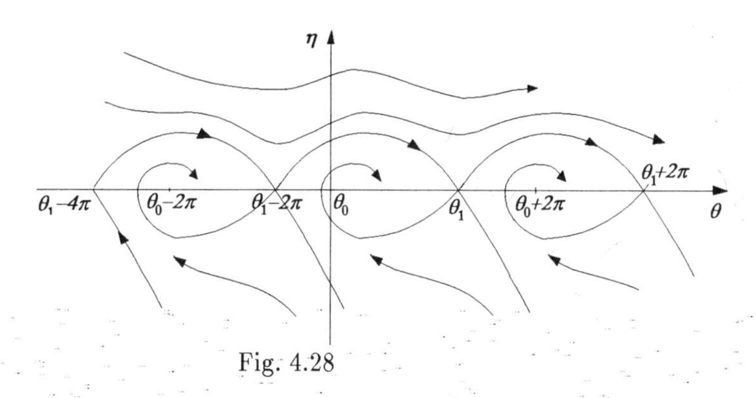

Fig. 4.28

3) The whole phase space is partitioned into the domains of attraction of stable equilibria. The boundaries of these domains are separatrices of saddle equilibria (Fig. 4.29).

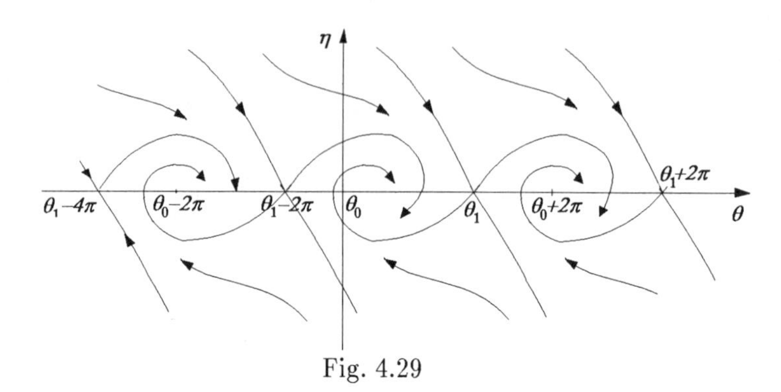

Fig. 4.29

Thus, the determination of domains of attraction of stable equilibria and the whole global analysis of system (4.22) are reduced to a computation or an estimation of one trajectory only, namely a separatrix $\tilde{\tilde{\eta}}(t)$, $\tilde{\tilde{\theta}}(t)$. In case 1) it is also important that a separatrix $\tilde{\eta}(t)$, $\tilde{\theta}(t)$ can be computed or estimated.

F. Tricomi and his followers have obtained various analytic estimates of these separatrices. Case 2) in the parameter space $\{\alpha, \gamma\}$ corresponds to interchanging the qualitative pictures of trajectory dispositions (i.e., corresponds to the replacement of case 1) by case 3) or vice versa). Such a qualitative change is called a bifurcation and the parameters α and γ, corresponding to case 2), are called parameters of bifurcation. The evaluations of separatrices lead to the evaluations of parameters of bifurcations.

In the 60–70s of last century for the determination of separatrices and, consequently, of parameters of bifurcation the numerical methods were applied [40].

The computation of separatrices is divided by two parts. In some sufficiently small neighborhood of the saddle point the separatrix is approximated by a linear function with the coefficient

$$-\frac{\alpha}{2} \pm \sqrt{\frac{\alpha^2}{4} - \varphi'(\theta_1)}\ .$$

Outside this neighborhood some numerical method for a solution of ordinary differential equations is used (the most often it is the Runge—Kutta method).

By using the numerical methods, mentioned above, the bifurcation curve is computed (see Fig. 4.30).

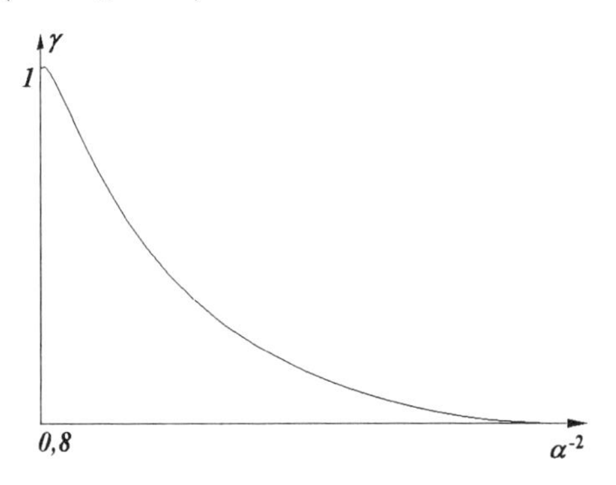

Fig. 4.30

The behavior of trajectories in the phase space, shown in Fig. 4.28, corresponds to the points on this curve. A structure, shown in Fig. 4.29, corresponds to the points, placed under this curve. A situation, shown in Fig. 4.27, corresponds to the points, placed above this curve and under the straight line $\{\gamma = 1\}$. Figure 4.20 corresponds to the points, placed above the straight line $\{\gamma = 1\}$.

The above imply the following statement.

Lemma 4.2. *Any solution $\eta(t)$, $\theta(t)$ with initial data from the domain*

$$\left\{\frac{1}{2}\eta^2 + \int_{\theta_1}^{\theta} \varphi(\theta)\,d\theta < 0, \ \ \theta \in (\theta_1 - 2\pi, \theta_1)\right\}$$

tends to equilibrium $\eta = 0$, $\theta = \theta_0$ as $t \to +\infty$.

Now we discuss the problem on a limit load of synchronous machine.

The limit load problem. Consider a synchronous electromotor, which rotates the rolls of hot rolling mill (Fig. 4.31). Till a red-hot bar moves along lower rolls only, we may assume that a synchronous machine load is equal to zero and the machine operates in a synchronous regime, i.e., $\theta(t) \equiv 0$, $\dot{\theta}(t) \equiv 0$. At time $t = \tau$ the bar enters the part of rolling mill, in which a rolling occurs. The bar moves forward and decreases in thickness due to the upper and lower rolls, rotating in opposite directions. Obviously, at time $t = \tau$ a load-on occurs.

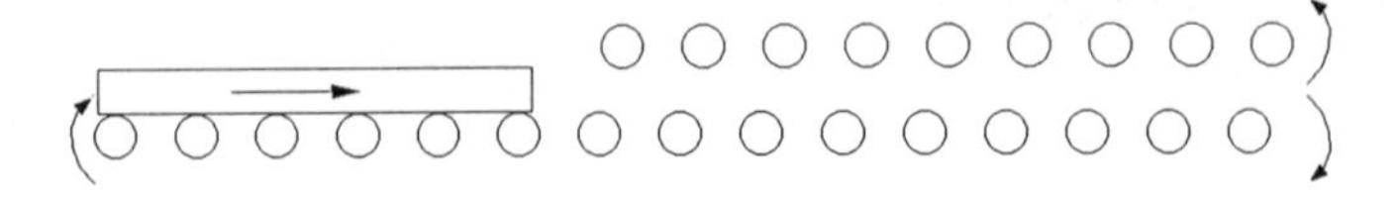

Fig. 4.31

Thus, for $t < \tau$ the process is described by the following equation

$$\ddot{\theta} + \alpha\dot{\theta} + \sin\theta = 0 \tag{4.32}$$

and corresponds to the solution $\theta(t) \equiv 0$, $\dot{\theta}(t) \equiv 0$. For $t > \tau$ we have

$$\ddot{\theta} + \alpha\dot{\theta} + \sin\theta = \gamma \tag{4.33}$$

and the process corresponds to a solution with initial data $\theta(\tau) = 0$, $\dot{\theta}(\tau) = 0$.

We observe that such a solution of equation (4.33) is not equilibrium and the problem lies in the fact that after a load-on the synchronous machine remained synchronous as before. In other words, it is necessary for solution of equation (4.33) with initial data $\theta(\tau) = 0$, $\dot{\theta}(\tau) = 0$ to be in a domain of attraction of a new stable equilibrium $\theta(t) \equiv \theta_0$, $\dot{\theta}(t) \equiv 0$.

To solve this problem we apply Lemma 4.2, which can be restated in the following way.

Theorem 4.2. *An admissible load-on γ is a value γ satisfying the following inequality*

$$\int_{\theta_1}^{0} (\sin\theta - \gamma)\, d\theta < 0. \tag{4.34}$$

Theorem 4.2 implies that the limit load-on $\gamma = \Gamma$ can be obtained from the equality of areas shown in Fig. 4.32

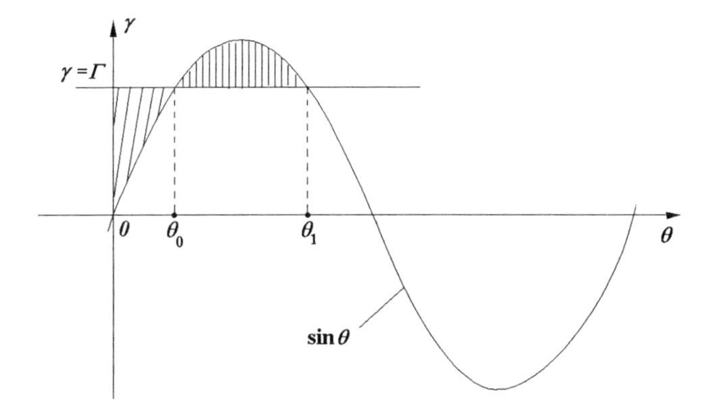

Fig. 4.32

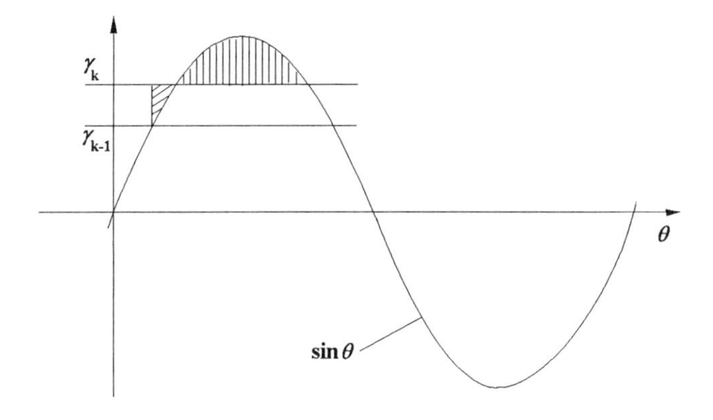

Fig. 4.33

In the engineering practice such a method to determine the limit load-on is called a method of areas.

Now we discuss what advantage has an algorithm of a slow loading of a synchronous machine. Suppose, we have m steps of loading. At the k−th step at time $\tau_k > \tau_{k-1}$ the load increases instantly from γ_{k-1} to γ_k. Let a difference $\tau_k - \tau_{k-1}$ be sufficiently large, namely such that a solution with initial data $\theta(\tau_k)$, $\dot{\theta}(\tau_k)$ (the previous equilibrium) is in a sufficiently small neighborhood of a new equilibrium. In this case by Lemma 4.2 the initial data $\theta(\tau_k)$, $\dot{\theta}(\tau_k)$ belong to a domain of attraction of a new stable equilibrium if the upper shaded area in Fig. 4.33 is greater than the lower shaded area. By choosing γ_{k-1} and γ_k sufficiently close together, we can always satisfy this condition.

Thus, a synchronous machine can be slow loaded up to any value $\gamma < 1$.

Note that the start of synchronous generators and their loading are complicated dynamic processes taking sometimes several tens of minutes.

Note also that at present there are many synchronous machine control systems, which make use of an exciting winding. The values $\theta(t)$ and $\dot{\theta}(t)$ are transformed into an additional voltage, which is delivered to an exciting winding. This voltage creates a controlled force moment acting on a rotor and refining the dynamic properties of synchronous machine. We make mention here of the book of M.M. Botvinnik [8] in which such a control is considered. M.M. Botvinnik is also far-famed as a multiple world chess champion.

Phase locked loops. The synchronization principle is the central one in transmitting video pictures. We discuss a basic principle of black-and-white display transmission.

At broadcast station, on a photomatrix there is a picture of an object. An electronic ray passes sequentially along the lines of this matrix and, closing a circuit at time t, generates a voltage $u(t)$, depending on brightness of a matrix element in which at time t the electronic ray occurs (Fig. 4.34).

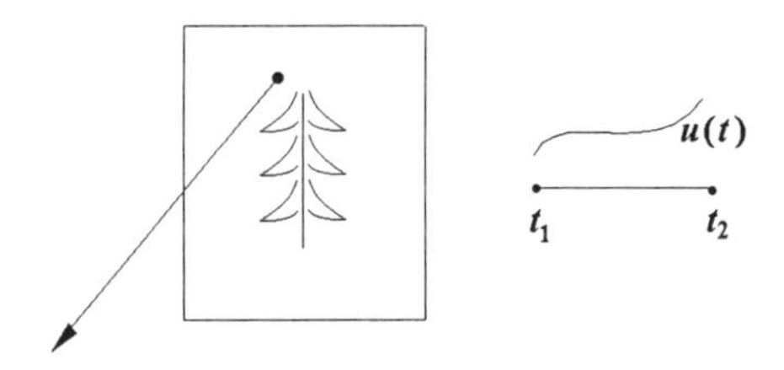

Fig. 4.34

A scanning of the photomatrix takes the time interval $[t_1, t_2]$. Then a high-frequency modulation of the signal $u(t)$ occurs which is transmitted via a space to a television receiver. Here a demodulation arises and the signal is transmitted to a driver circuit of an electronic ray in a television electron-beam tube. The greater is the value of $u(t)$, the greater is the intensity of electronic ray and, consequently, the greater is the brightness of the corresponding point of the display, at which the electronic ray occurs. The ray, passing through a deflector, scans sequentially display lines and creates a picture on the display.

It is clear that in order to avoid the transmission errors, the generators of deflection of electronic rays at the television station and in the television receiver must work synchronously. In other words, the frequencies of both generators must coincide. This is an objective of the control systems being phase locked loops. These systems work in the following way.

Information on the frequency ω_0 of generator at the television station is transmitted to each television receiver by means of a sync-pulse signal, which is a part of television signal (Fig. 4.35).

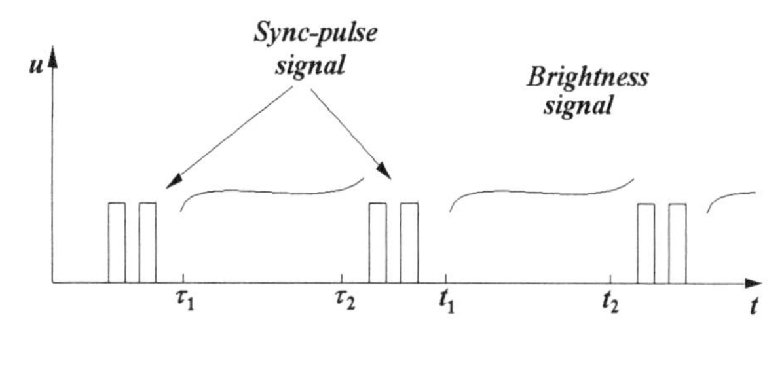

Fig. 4.35

In television receivers this signal is separated and information on ω_0 enters into the phase locked loops, shown in Fig. 4.36.

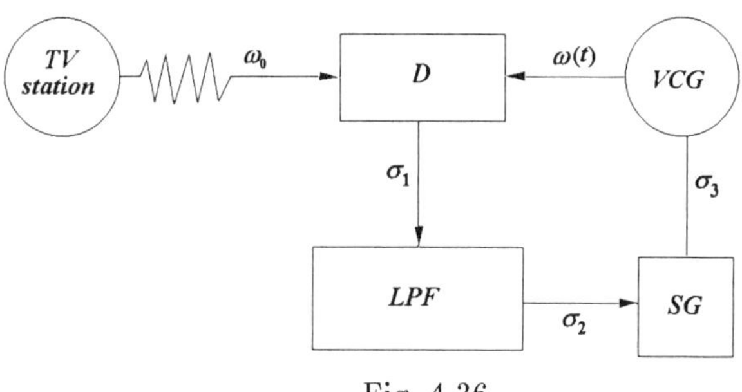

Fig. 4.36

In the case of open loop when the steering gear (SG) does not act on a voltage controlled generator (VCG), the latter (VCG) works in such a way that frequency ω_1 is kept constant. We also assume that the frequency ω_0 of a TV station generator is constant.

The objective of a control is to eliminate the frequency misalignment $\Gamma = \omega_0 - \omega_1$. After initiating the control system, shown in Fig. 4.36 at time $t = 0$, the value ω is not constant: $\omega = \omega(t)$, $\omega(0) = \omega_1$. The objective of a control system is the following synchronization: $\omega(t) \to \omega_0$ as $t \to +\infty$. The functions

$$\theta_0(t) = \theta_0(0) + \omega_0 t, \quad \theta(t) = \theta(0) + \int_0^t \omega(\tau)\, d\tau \tag{4.35}$$

are called phases of generators. The numbers $\theta_0(0)$ and $\theta(0)$ are the values of these phases at time $t = 0$.

The phases $\theta_0(t)$ and $\theta(t)$ are inputs of the block $\mathcal{D}$, which is called a phase detector. There are many different electronic schemes being phase detectors. Their description can be found in [31, 40] It is significant that in all these systems the output of block $\mathcal{D}$ has the form

$$\sigma_1 = F(\theta_0 - \theta). \tag{4.36}$$

Here the function $F(x)$ is 2π-periodic. Typical functions $F(x)$ are $\sin x$ and the functions whose graphs are shown in Fig. 4.37 and 4.38.

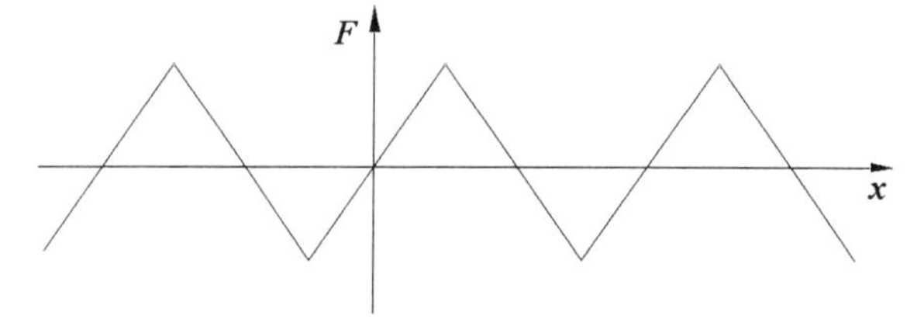

Fig. 4.37

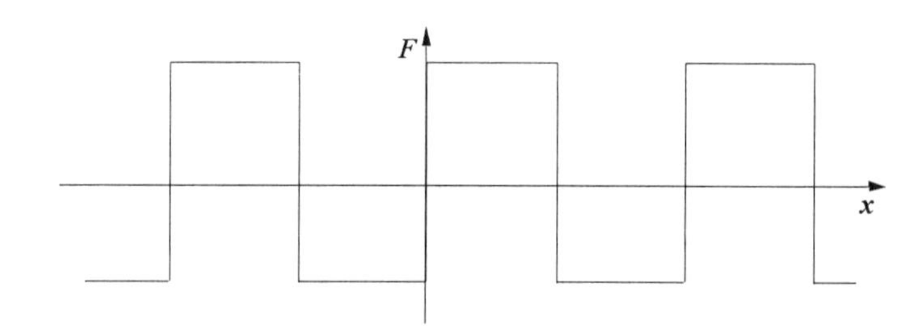

Fig. 4.38

The signal $\sigma_1(t) = F(\theta_0(t) - \theta(t))$ enters the input of a low pass filter (LPF), which suppress disturbances arising in a process of transmitting information on ω_0. In Chapter 2 RC- and RLC-circuits are considered which are often used as filters. In the general case a low pass filter is a linear block with an input σ_1, an output σ_2, and a transfer function $K(p)$.

The characteristic of a steering gear is usually linear. The change of frequency of a voltage controlled generator under the action of a signal of steering gear can be given by

$$\omega(t) = \omega_1 + \sigma_3(t), \qquad \sigma_3(t) = S\,\sigma_2(t), \tag{4.37}$$

where S is a number. The action of the most of steering gears (SG) is based on the change of reactivity of the contour of a voltage controlled generator. The outlines of such devices can be found in the above-mentioned books [31, 40].

In the simplest case that a low pass filter does not enter into system (4.35)—(4.37), we have the following differential equation of the first order, describing the phase locked loops,

$$(\theta_0 - \theta)^{\bullet} = \omega_0 - \omega_1 - SF(\theta_0 - \theta).$$

Using notation $\sigma = \theta_0 - \theta$ and recalling that $\Gamma = \omega_0 - \omega_1$, we obtain finally the equation

$$\frac{d\sigma}{dt} + SF(\sigma) = \Gamma. \tag{4.38}$$

Consider now the equations of the phase locked loop with RC-filter.

Since for RC-circuit the signals σ_1 and σ_2 are related by an equation

$$RC\,\frac{d\sigma_2}{dt} + \sigma_2 = \sigma_1,$$

we obtain, taking into account equations (4.35)–(4.37), the following equations, describing the work of phase locked loop:

$$\begin{aligned} &RC\,\frac{d\sigma_2}{dt} + \sigma_2 = F(\sigma), \\ &\dot{\sigma} = \omega_0 - \omega_1 - S\sigma_2. \end{aligned} \tag{4.39}$$

Using a relation $\omega_0 - \omega_1 = \Gamma$ and substituting the expression for $\sigma_2(t)$, obtained from the second equation of system (4.39), into the first equation, we obtain

$$RC\,\frac{d^2\sigma}{dt^2} + \frac{d\sigma}{dt} + SF(\sigma) = \Gamma, \tag{4.40}$$

where $\sigma = \theta_0 - \theta$. If a characteristic of the phase detector is given by $F(\sigma) = \sin\sigma$, then from equation (4.40) we obtain again equation (4.21)

$$\ddot{\sigma} + \alpha\dot{\sigma} + b\sin\sigma = \gamma,$$

where $\alpha = 1/(RC)$, $b = S/(RC)$, $\gamma = \Gamma/(RC)$.

The phase portrait, shown in Fig. 4.29, corresponds to a solution of the control problem of setting a synchronous process of generator.

Further development of mathematical tools for the study of the phase locked loops with more complicated filters and characteristics of detectors can be found in the books [28, 29].

Finally we note that all the control systems, considered in §1 and §2, have a cylindrical phase space. Consider such a space.

From Proposition 4.1 §1 it follows that the properties of the systems are invariant with respect to the shift $x + d$, where

$$x = \begin{pmatrix} \eta \\ \sigma \end{pmatrix}, \qquad d = \begin{pmatrix} 0 \\ 2\pi \end{pmatrix},$$

We see that if $x(t)$ is a solution of the system, then $x(t) + d$ is also solution of it.

Consider a discrete group

$$\Gamma = \{x = kd \mid \ k \in \mathbb{Z}\}.$$

Here $\mathbb{Z}$ is a set of integer numbers.

Let us consider now a factor-group $\mathbb{R}^n/\Gamma$ whose elements are residue classes $[x] \in \mathbb{R}^n/\Gamma$, which are defined by

$$[x] = \{x + u \mid u \in \Gamma\}.$$

We introduce the so-called plane metric

$$\rho([x],[y]) = \inf_{\substack{z\in[x]\\ \vartheta\in[y]}} |z - \vartheta|.$$

Here, as previously, $|\cdot|$ is Euclidean norm in $\mathbb{R}^n$.

It follows from Proposition 4.1 §1 that $[x(t)]$ is a solution and the metric space $\mathbb{R}^n/\Gamma$ is a phase space. It is partitioned into nonintersecting trajectories $[x(t)],\ t \in \mathbb{R}^1$.

We can find the following diffeomorphism between $\mathbb{R}^n/\Gamma$ and the surface of cylinder $\mathbb{R}^1 \times C$. Here C is a circle of radius 1. Consider a set $\Omega = \{x|\ \eta \in \mathbb{R}^1, \sigma \in (0, 2\pi]\}$, which contains exactly one representative of each class $[x] \in \mathbb{R}^n/\Gamma$. Let us cover the surface of the cylinder by the set Ω (see Fig.39).

The mapping so constructed is a diffeomorphism. The surface of cylinder is also partitioned into nonintersecting trajectories. Such a phase space is called cylindrical.

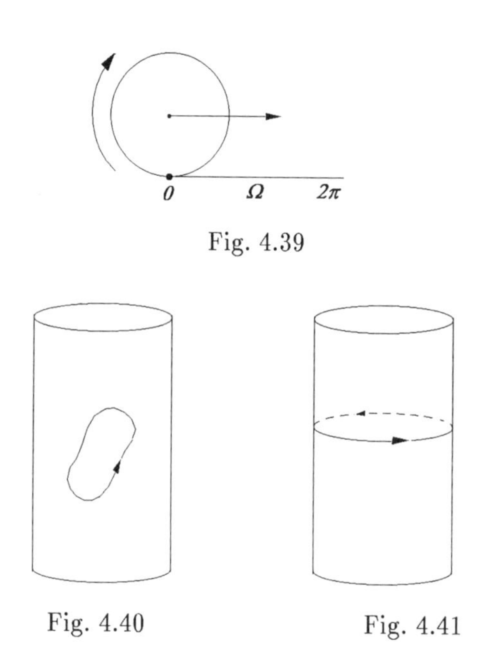

Fig. 4.39

Fig. 4.40

Fig. 4.41

Sometimes such a space is more convenient than $\mathbb{R}^2$. Really, in cylindrical phase space there does not exist a multiplicity: all the values $\sigma + 2k\pi$ correspond to one state of system only. Besides in this case there are two types of the closed trajectories, which correspond to periodic processes of control systems. The closed trajectories of the first type can homotopically be collapsed onto a point (Fig. 4.40). This closedness is preserved in passing to $\mathbb{R}^2$. The closed trajectories of the second type, which encircle the cylinder, cannot be collapsed homotopically onto a point. Such a closedness vanishes when passing to $\mathbb{R}^2$ (Fig. 4.41).

4.3 The mathematical theory of populations

A biological population, having an environment facilities such as a plenty of food, an absence of enemies, and a large livingspace, reproduces themselves in direct proportion to its strength. Let the number of persons be sufficiently large. Keeping in mind a certain idealization, assume that $n(t)$ is a smooth function of t. This reproduction is described by the following differential equation

$$\frac{dn}{dt} = \lambda n. \tag{4.41}$$

Here a constant positive number λ is a birthrate of population. Equation (4.41) can also be found from the analysis of economic growth, assuming that on large time intervals the growth coefficient λ is constant. Note that on a large time interval a bank deposit rate λ is also described by equation (4.41).

We consider the solution of this equation in the following form

$$n(t) = e^{\lambda t} n(0). \tag{4.42}$$

Under the ideal conditions the populations obey the law of exponential increase (4.42), which was experimentally observed in the case of the bacteria reproduction.

However in ecological systems we see a strong competition of populations, concerning with space and food, and predators annihilating preys. Besides the human interference may also depress some populations. Thus, in the ecological systems there exist many feedbacks and therefore the human attempts to control an ecological system (for example, by means of shooting predators) may have also objectionable results.

The study of the mathematical models of population interactions was begun in 20th of the previous century and in 1931 the notorious book of Vito Volterra [47], devoted to this problem, has been published.

We consider the Lotka-Volterra equations, which describe the interaction of two species: predators and preys. This mathematical model, its various generalizations, and the other of interesting population models are discussed in the book [47].

The preys (hare, for example) are assumed to be reproduced themselves under the absence of predators (wolves, for example) according to mathematical model (4.41). In the case that the preys (a food of predators) are absent the predators die out, what corresponds to a negative growth of the coefficient λ.

Assuming the coexistence of these two species, the predators and preys can encounter to one another. The number of these encounters is in direct proportion to strength of predators and preys $\alpha n_1(t)n_2(t)$. These encounters decrease a population of preys (with a coefficient $-\beta_1$) and increase a population of predators (with a coefficient β_2).

Arguing as above, we arrive at the following differential equations for the strengths of the prey population $n_1(t)$ and the predators population $n_2(t)$

$$\begin{aligned} \dot{n}_1 &= \lambda_1 n_1 - \beta_1 \alpha n_1 n_2, \\ \dot{n}_2 &= -\lambda_2 n_2 + \beta_2 \alpha n_1 n_2. \end{aligned} \tag{4.43}$$

Here n_1 and n_2 are nonnegative. In this case system (4.43) has the following solutions

$$\begin{aligned} n_1(t) &\equiv 0, \qquad n_2(t) = e^{-\lambda_2 t} n_2(0), \\ n_2(t) &\equiv 0, \qquad n_1(t) = e^{\lambda_1 t} n_2(0), \\ n_1(t) &\equiv n_2(t) \equiv 0. \end{aligned}$$

Thus, the phase space of system (4.43) is the first quadrant of the plane $\{n_1 \geq 0,\ n_2 \geq 0\}$ (Fig. 4.42)

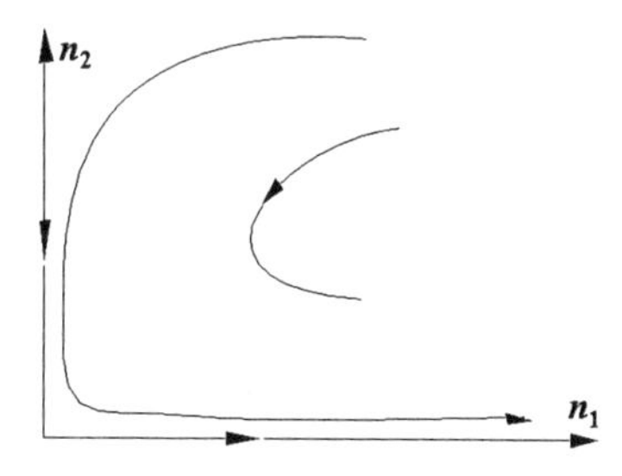

Fig. 4.42

Now we show that for $n_1 > 0$, $n_2 > 0$ system (4.43) has the first integral

$$V(n_1, n_2) = \lambda_2 \ln n_1 - \alpha\beta_2 n_1 + \lambda_1 \ln n_2 - \alpha\beta_1 n_2.$$

Really,

$$\begin{aligned}\dot V(n_1(t), n_2(t)) &= \lambda_2 \frac{\dot n_1}{n_1} - \alpha\beta_2 \dot n_1 + \lambda_1 \frac{\dot n_2}{n_2} - \alpha\beta_1 \dot n_2 = \\ &= \lambda_2(\lambda_1 - \alpha\beta_1 n_2) - \alpha\beta_2(\lambda_1 n_1 - \alpha\beta_1 n_1 n_2) + \\ &+ \lambda_1(-\lambda_2 + \alpha\beta_2 n_1) - \alpha\beta_1(-\lambda_2 n_2 + \alpha\beta_2 n_1 n_2) = 0.\end{aligned}$$

Whence it follows that the function

$$W(n_1, n_2) = e^{V(n_1,n_2)} = g(n_1)\, h(n_2),$$

where

$$g(n_1) = n_1^{\lambda_2} e^{-\alpha\beta_2 n_1}, \qquad h(n_2) = n_2^{\lambda_1}\, e^{-\alpha\beta_1 n_2},$$

is also the first integral of system (4.43).

The functions g and h have the maximums at the points

$$n_1 = \frac{\lambda_2}{\alpha\beta_2}, \qquad n_2 = \frac{\lambda_1}{\alpha\beta_1} \tag{4.44}$$

respectively. The graphs of these functions are shown in Fig. 4.43.

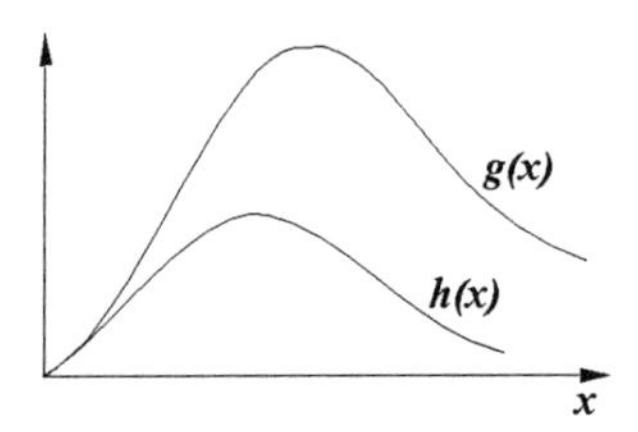

Fig. 4.43

The function $W(n_1, n_2)$ has a property

$$W(n_1, n_2) \le W\left(\frac{\lambda_2}{\alpha\beta_2}, \frac{\lambda_1}{\alpha\beta_1}\right).$$

Whence it follows that the level curves of function $W(n_1, n_2)$ are the closed curves encircling a point given by (4.44).

Thus the trajectories of system (4.43) are placed on the closed level curves and point given by (4.44) is an equilibrium state (see Fig. 4.44).

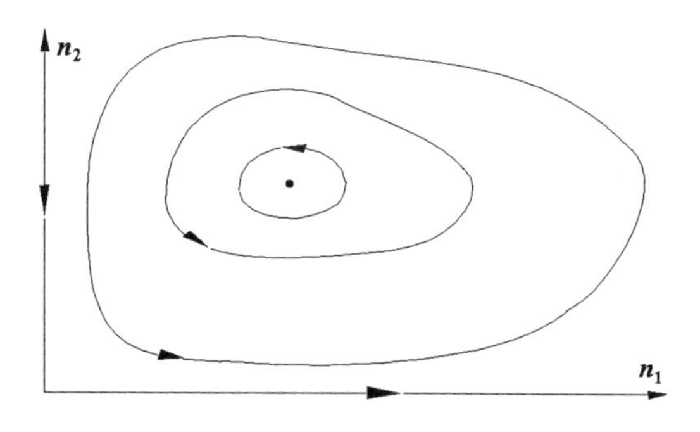

Fig. 4.44

The strengths $n_1(t)$ and $n_2(t)$ of preys and predators respectively, as it can be seen from the phase portrait in Fig. 4.44, are periodic functions. Such an oscillating character of the change of a strength of population one can observe in various ecological systems and the Lotka-Volterra equations describe qualitatively the interaction of populations of predators and preys to one another in a sufficiently adequate manner.

Consider system (4.43) in the form

$$(\ln n_1)^{\bullet} = \lambda_1 - \beta_1 \alpha n_2,$$
$$(\ln n_2)^{\bullet} = -\lambda_2 + \beta_2 \alpha n_1$$

and integrate the left-hand and right-hand sides of these equations from 0 to T, where T is a period of solution $n_1(t)$, $n_2(t)$. Then we obtain

$$\frac{1}{T}\int_0^T n_2(t)\,dt = \frac{\lambda_1}{\alpha\beta_1}, \tag{4.45}$$

$$\frac{1}{T}\int_0^T n_1(t)\,dt = \frac{\lambda_2}{\alpha\beta_2}. \tag{4.46}$$

Recall that a magnitude

$$\frac{1}{T}\int_0^T x(t)\,dt$$

is called a mean value of a T-periodic function $x(t)$.

Thus, we proved the following

Theorem 4.3. *The mean values of functions $n_1(t)$ and $n_2(t)$ do not depend on the initial data $n_1(0)$ and $n_2(0)$ respectively and coincide with the stationary values* (4.44).

Consider now the case that strengths of populations are controlled by means of an annihilation of persons of each species. This annihilation is assumed to be in the direct proportion to the strength of population. In this case the equations of interaction of populations have the form

$$\begin{aligned} \dot{n}_1 &= (\lambda_1 - \gamma_1)n_1 - \beta_1 \alpha n_1 n_2, \\ \dot{n}_2 &= -(\lambda_2 + \gamma_2)n_2 + \beta_2 \alpha n_1 n_2. \end{aligned} \tag{4.47}$$

Here γ_1 and γ_2 are intensity coefficients of annihilation.

We consider first the case $\lambda_1 < \gamma_1$ (it means that the annihilation of preys is more intensive than their reproduction in the absence of predators). In this case both populations die out.

Really, consider a function $V(n_1, n_2) = \beta_2 n_1 + \beta_1 n_2$. For solutions $n_1(t)$ and $n_2(t)$ of equation (4.47) we have

$$\begin{aligned} &\dot{V}(n_1(t), n_2(t)) = (\lambda_1 - \gamma_1)\beta_2 n_1(t) - (\lambda_2 + \gamma_2)\beta_1 n_2(t) \le \\ &\le -\varepsilon V(n_1(t), n_2(t)), \end{aligned} \tag{4.48}$$

where $\varepsilon = \min(\gamma_1 - \lambda_1, \lambda_2 + \gamma_2)$. Considering inequality (4.48) in the following form

$$\left(V(n_1(t), n_2(t))e^{\varepsilon t}\right)^{\bullet} \le 0 \qquad \forall\, t \ge 0 \tag{4.49}$$

and integrating both sides of inequality (4.49) from 0 to t, we obtain an estimate

$$V(n_1(t), n_2(t)) \le e^{-\varepsilon t} V(n_1(0), n_2(0)). \tag{4.50}$$

By (4.50)

$$\lim_{t \to +\infty} n_1(t) = 0, \qquad \lim_{t \to +\infty} n_2(t) = 0.$$

From the last relations a complete annihilation of preys and an extinction of predators follows.

Now we consider the most interesting case $\lambda_1 > \gamma_1$.

In this case in place of formulas (4.45) and (4.46) we obtain

$$\frac{1}{T}\int_0^T n_2(t)\,dt = \frac{\lambda_1 - \gamma_1}{\alpha\beta_1},$$

$$\frac{1}{T}\int_0^T n_1(t)\,dt = \frac{\lambda_2 + \gamma_2}{\alpha\beta_2}.$$

These relations can be restated in the following assertion.

Theorem 4.4 *If two species are exterminated in direct proportion to the number of persons, then the mean value of preys increases and the mean value of predators decreases.*

A nontrivial fact is that, *shooting hare and do not hunting wolves, we do not affect a mean strength of hare. In this case the mean strength of wolves decreases.*

Chapter 5

Discrete systems

5.1 Motivation

1. Mathematical motivation. A generator of Cantor set.

In the previous chapter a wide variety of different trajectories in two-dimensional phase spaces was demonstrated. These are trajectories, tending to equilibrium or to infinity, the trajectories that are closed curves corresponding to periodic solutions or are closed in a two-dimensional cylindrical phase space.

In a one-dimensional phase space the trajectories of a differential equation

$$\frac{dx}{dt} = f(x), \qquad x \in \mathbb{R}^1 \tag{5.1}$$

with the continuous right-hand side have no such a variety. From the fact that trajectories of equation (5.1) either fill the whole intervals in $\mathbb{R}^1$ or turn out to be stationary points, coinciding with the zeroes of the function $f(x)$, we can conclude that any solution of equation (5.1) as $t \to +\infty$ tends either to stationary point or to infinity (Fig. 5.1).

x

Fig. 5.1

A situation, which is rather different from the above, occurs in the case of one-dimensional discrete equation

$$x_k = f(x_{k-1}). \tag{5.2}$$

This equation is a recurrent formula to determine a sequence x_k, $k = 1, 2, \ldots$. To find uniquely the sequence x_k the initial data x_0 must be given. This case is similar to that of solving the Cauchy problem for differential equations.

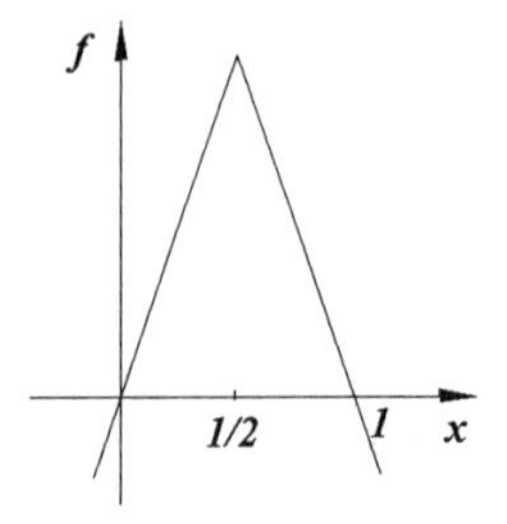

Fig. 5.2

Let the function $f(x)$ in equation (5.2) be given in the following form (Fig. 5.2)

$$f(x) = \begin{cases} 3x & \text{for} \quad x < 1/2, \\ 3(1-x) & \text{for} \quad x \geq 1/2. \end{cases} \tag{5.3}$$

Then equation (5.2) has exactly two stationary solutions

$$x_k \equiv 0, \qquad x_k \equiv 3/4.$$

Lemma 5.1. *For any $x_0 \overline{\in} [0, 1]$ the solution x_k tends to $-\infty$ as $k \to \infty$.*

Proof. For $x_0 < 0$ from (5.3) it follows that $x_k < 0$ for all $k = 0, 1, \ldots$ and

$$x_k = 3^k x_0. \tag{5.4}$$

For $x_0 > 1$ we obtain $x_1 < 0$. Consequently,

$$x_k = 3^{k-1} x_1. \tag{5.5}$$

The statement of Lemma 5.1 follows directly from relations (5.4) and (5.5).

Now we discuss the map f of the segment $[0, 1]$. Let us assign the segment $[0, 1]$ to each of two segments $[0, 1/3]$ and $[2/3, 1]$. The interval $(1, 3/2)$ is assigned to the interval $(1/3, 2/3)$. By Lemma 5.1 all the solutions x_k of equation (5.2) with initial data from interval $(1/3, 2/3)$ tend to infinity as $k \to \infty$. Thus, interval $(1/3, 2/3)$ is excluded from the segment $[0, 1]$.

The map $f(f(\cdot))$ takes each of the segments $[0,1/9]$, $[2/9,3/9]$, $[6/9,7/9]$, $[8/9,1]$ to the segment $[0,1]$. The solutions x_k with initial data from intervals $(1/9,2/9)$, $(1/3,2/3)$, $(7/9,8/9)$ are excluded from the segment $[0,1]$ and by Lemma 5.1 it tend to infinity as $k \to +\infty$.

Solutions x_N belong to the segment $[0,1]$ if and only if the initial data are in the segments

$$\left[0,\ \frac{1}{3^N}\right], \quad \left[\frac{2}{3^N},\ \frac{3}{3^N}\right], \quad \left[\frac{6}{3^N},\ \frac{7}{3^N}\right], \ldots, \left[\frac{3^N-1}{3^N},\ 1\right].$$

The process of eliminating the medial parts of remaining segments is usually illustrated by the following figure (Fig. 5.3).

Fig. 5.3

The part of the segment $[0,1]$ that remains after infinitely many of procedures of deleting the medial parts of remaining segments is called a Cantor set.

It is a very "holey" set. In any neighborhood of a point of the set there are holes being small intervals, which do not belong to this set.

Can one measure this set? We shall answer to this question in the frame of the measure theory.

At first we make use of the Lebesgue measure. In this case it is sufficient to measure the final lengths of segments after the N–th iteration, to summarize them, and to tend N to infinity. This implies that

$$\sum_{j=0}^{2^N} l_j = 2^N \frac{1}{3^N} \underset{N\to\infty}{\to} 0.$$

Thus, the Lebesgue measure of a Cantor set equals zero.

In 1916 F. Hausdorff has introduced another definition of measure (to be precise, of an outer measure).

Let us consider the Hausdorff measure of the Cantor set. We measure the set by not linear meters m, square meters m^2 or cubic meters m^3 but by m^d, where d is any positive number.

In this case for each segment, obtained after N iterations, we find the following measure

$$\left(\frac{1}{3^N}\right)^d.$$

We see that if this segment is assumed to be measured in square meters, then in terms of outer measurement (outer measure), covering the segment by the square with side $1/3^N$ (Fig. 5.4), we obtain the following "outer" value $(1/3^N)^2$. Extending this expression to arbitrary d, we obtain a value $(1/3^N)^d$.

Summing over the rest of segments, we find

$$2^N\left(\frac{1}{3^N}\right)^d = \left(\frac{2}{3^d}\right)^N.$$

This value tends neither to zero nor to infinity in the only case

$$d = \frac{\log 2}{\log 3}.$$

Thus, the Hausdorff dimension of a Cantor set is equal to $\log 2/\log 3$ and its Hausdorff measure is one.

The existence of sets similar to the Cantor ones stimulated the development of the modern theory of measure and a metric dimension of sets.

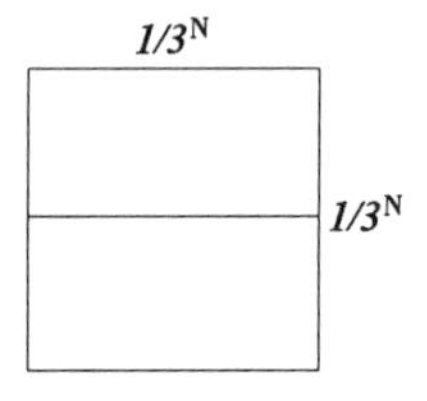

Fig. 5.4

Note that if the initial point x_0 is in a Cantor set K, then the corresponding solution x_k belongs to K for all k. The set K, possessing such a property, is called invariant (or positively invariant).

In the modern theory of dynamical systems the invariant sets of a non-integer Hausdorff dimension are called strange. At present the continuous and discrete systems, having such invariant sets, are intensively studied. In discrete systems such sets exist in the simplest one-dimensional models.

For continuous system such sets are found due to computer experiments for phase spaces of dimension greater than two. (See [42].)

The generator considered above tends to infinity all the points from $\mathbb{R}^1 \setminus K$.

We consider now the discrete system (5.2) with the phase space $[-1, 3/2]$ and with the continuous function $f(x)$ satisfying the following conditions

$$f(x) = \begin{cases} 3x & \text{for} \quad x \in [0, 1/2], \\ 3(1-x) & \text{for} \quad x \in [1/2, 1], \end{cases}$$

$$\begin{array}{ll} -1 \le f(x) \le 0 & \text{for} \quad x \in [1, 3/2], \\ \alpha x < f(x) \le 0 & \text{for} \quad x \in [-1, 0], \end{array}$$

where α is a number from the interval $(0, 1)$.

Lemma 5.2. *For any $x_0 \in [-1, 0] \cup [1, 3/2]$ the solution x_k tends to zero as $k \to +\infty$.*

Proof. Put $x_0 \in [1, 3/2]$. Then $x_1 \in [-1, 0]$. The set $[-1, 0]$ is positively invariant, i.e., for $x_0 \in [-1, 0]$ we have $x_k \in [-1, 0]$ $\forall 1, 2, \ldots$, in which case

$$x_k > \alpha^{k-1} x_1 \to 0$$

as $k \to +\infty$. This completes the proof of Lemma 5.2.

Using Lemma 5.2 and arguing as in the case of constructing the Cantor set, we see that in addition to the invariance of K with respect to the map f there exists a property of a global attraction, namely any solution x_k in the phase space $[-1, 3/2]$ tends to set K as $k \to +\infty$:

$$\rho(x_k, K) = \inf_{z \in K} |z - x_k| \to 0 \tag{5.6}$$

as $k \to +\infty$.

The invariant bounded sets, possessing attracting property (5.6), are called attractors. The attractors, having the noninteger Hausdorff dimensions, are called strange attractors. In some works an attractor is said to be "strange" if it has a property of inner instability, which is in the following: for small changes of initial data $x_0 \in K$, the difference between the solutions, corresponding to these initial data, is not small for sufficiently large values of k.

For the Cantor set generator such a sensitivity with respect to initial data from K occurs.

Note that the numerical methods for solving differential equations, which are based on the idea of discretization, lead to discrete equations (5.2) with an n-dimension phase space $\mathbb{R}^n$.

2. Examples from technology.

We now pass from the Cantor set to the analysis of a work of production storage. Consider a storage for n different component parts of a product.

At the end of each day one makes the report on the numbers of the component parts. These reports can be regarded as discrete times: $t = 0, 1, 2, \ldots, k$. The components of vector x_k are the numbers of component parts of the similar type, which are stocked at the end of previous day. Each of components of vector u_k is a number of component parts of a certain type, which are delivered at the storage for a day on account of external supplies.

Suppose that for the products to be fabricated it is necessary to take $\mathcal{D}x_k$ component parts from the storage for a day. Here $\mathcal{D}$ is a diagonal matrix.

It follows that industrial chain: a storage – an assembly plant can be described by the following discrete equations

$$x_{k+1} = x_k - \mathcal{D}x_k + u_k, \quad \sigma_k = c^* x_k. \tag{5.7}$$

Here the values σ_k correspond to the finished goods in a day. The components of vector c coincide with the diagonal elements of matrix $\mathcal{D}$.

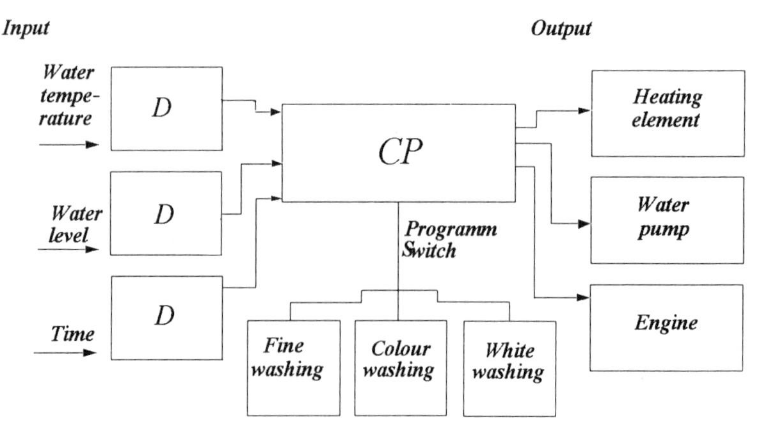

Fig. 5.5

As in Chapter 2, equation (5.7) can be regarded as a linear discrete block with the input u_k and the output σ_k. Introduction to the theory of such systems is presented in the next section.

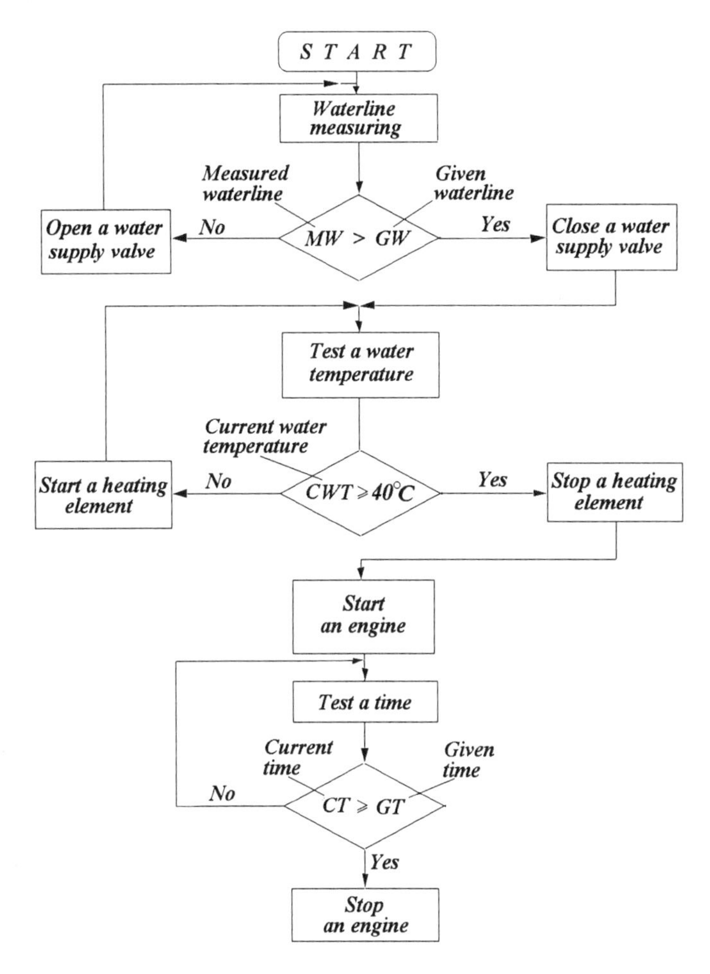

Fig. 5.6

The use of computers in the control systems permits us to describe the whole system or its part by discrete equations.

As an example we consider the use of microcomputers in a control system of a washing machine [10]. The block diagram of such a system is shown in Fig. 5.5.

The system controls three values: a temperature of water, a water level, and a washing time. By a switching element one of programs is started, which yields three parameters: a water level, a water temperature, and a washing time. By means of the sensors of a water level, a temperature, and a timer, prescribing the work time of electromotor, the current information enters the input in the form of continuous electric signals. A digitizer (D) transforms an analog information for a central processor (CP) in the discrete form. The signals of CP control heating elements, water pumps, and an electromotor. The signals of CP depend on a washing program. A flowchart of one of such programs is shown in Fig. 5.6.

Obviously, the system, described above, changes its states at discrete times, except for the input signals, which remain to be continuous until they are transformed by the digitizer.

5.2 Linear discrete systems

In this section we discuss a discrete analog of the linear theory, which was developed in Chapters 2 and 3 for continuous systems.

We consider first a linear discrete homogeneous system with a constant matrix

$$x_k = Ax_{k-1}, \qquad k = 0, 1, \dots \tag{5.8}$$

Here A is a constant $n \times n$-matrix, x is an element of an n-dimensional vector space.

In this case a solution of system (5.8) is given by

$$x_k = A^k x_0. \tag{5.9}$$

As in the theory of differential equations, we shall say that x_0 are initial data.

Sometimes it is useful to consider a similar matrix B

$$A = S^{-1}BS.$$

Then formula (5.9) takes the form

$$x_k = \underbrace{S^{-1}BSS^{-1}BSS^{-1}B\dots BS}_{k}\, x_0 = S^{-1}B^k Sx_0. \tag{5.10}$$

We recall that if the matrix B is a Jordan matrix, i.e.,

$$B = \begin{pmatrix} J_1 & & 0 \\ & \ddots & \\ 0 & & J_m \end{pmatrix},$$

where J_j are Jordan blocks, then the matrix B^k has a very simple form

$$B^k = \begin{pmatrix} J_1^k & & 0 \\ & \ddots & \\ 0 & & J_m^k \end{pmatrix},$$

where for the diagonal matrix J_j we obtain

$$J_j^k = \begin{pmatrix} \lambda_j^k & & 0 \\ & \ddots & \\ 0 & & \lambda_j^k \end{pmatrix}. \tag{5.11}$$

If a Jordan block is as follows

$$J_j = \begin{pmatrix} \lambda_j & 1 & & 0 \\ & \ddots & \ddots & \\ & & & 1 \\ 0 & & & \lambda_j \end{pmatrix}, \tag{5.12}$$

then

$$J_j^k = \begin{pmatrix} \lambda_j^k & \frac{k\lambda_j^{k-1}}{1!} & \frac{k(k-1)\lambda_j^{k-2}}{2!} & \cdots & \frac{k\ldots(k-l+2)\lambda_j^{k-l+1}}{(l-1)!} \\ & \ddots & \ddots & \ddots & \\ 0 & & & & \lambda_j^k \end{pmatrix}.$$

Here λ_j are the eigenvalues of the matrix A.

We assume further that x is an element of Euclidean vector space with the norm $|\cdot|$.

This implies the following

Theorem 5.1. *All the solutions of system* (5.8) *tend to zero as* $k \to +\infty$ *if and only if all the eigenvalues* λ_j *of the matrix* A *satisfy the inequality*

$$|\lambda_j| < 1. \tag{5.13}$$

For all the solutions of system (5.8) *to be bounded it is necessary and sufficient for the inequalities* $|\lambda_j| \leq 1$ *to be satisfied for the eigenvalues of the matrix* A, *corresponding to Jordan blocks of the form* (5.11), *and the inequalities* $|\lambda_j| < 1$ *are satisfied for the eigenvalues, corresponding to Jordan blocks of the form* (5.12).

Definition 5.1. *We shall say that a zero solution of system* (5.8) *is Lyapunov stable if for any* $\delta > 0$ *there exists a number* $\varepsilon > 0$ *such that from the inequality* $|x_0| \leq \delta$ *the relation* $|x_k| \leq \varepsilon$, $\forall\ k = 0, 1, \ldots$ *follows.*

Definition 5.2. *A zero solution of system* (5.8) *is said to be asymptotically stable if it is Lyapunov stable and all the solutions tend to zero as* $k \to +\infty$.

Let us remark that in this case by virtue of a linearity the two properties are equivalent: the property that the solutions with initial data from some small neighborhood of zero tend to zero and the property that all the solutions tend to zero as $k \to +\infty$.

Thus, condition (5.13) is necessary and sufficient for the asymptotic stability of a zero solution of system (5.8). We recall that for a system of differential equations

$$\frac{dx}{dt} = Ax$$

the necessary and sufficient condition of asymptotic stability is that the following inequality holds

$$\mathrm{Re}\,\lambda_j < 0$$

for all the eigenvalues λ_j of the matrix A.

We show now that, sometimes, for solving system (5.8) the change from formula (5.9) to (5.10) can be useful.

As an example, consider a discrete system of the form (5.8), which was probably the first system of this kind, studied by mathematicians. Early in the 13th century Fibonacci considered a discrete equation

$$f_k = f_{k-1} + f_{k-2}, \ k > 1, \qquad f_0 = f_1 = 1. \tag{5.14}$$

This equation generates the so-called Fibonacci numbers f_k. Let us find the explicit expressions for f_k, making use of the approach, described above.

We introduce the following notation

$$x_k = \begin{pmatrix} f_k \\ f_{k+1} \end{pmatrix}, \qquad x_0 = \begin{pmatrix} f_0 \\ f_1 \end{pmatrix} = \begin{pmatrix} 1 \\ 1 \end{pmatrix}.$$

Then equation (5.14) can be written as system (5.8) with a matrix

$$A = \begin{pmatrix} 0 & 1 \\ 1 & 1 \end{pmatrix}.$$

It is obvious that the Jordan form of matrix A is a matrix

$$B = \begin{pmatrix} p_1 & 0 \\ 0 & p_2 \end{pmatrix},$$

where

$$p_1 = \frac{1+\sqrt{5}}{2}, \qquad p_2 = \frac{1-\sqrt{5}}{2}.$$

To determine the matrix S, consider an equation

$$BS = SA, \tag{5.15}$$

which is a corollary of the equation $A = S^{-1}BS$.

Without loss of generality we can represent the matrix S in the following form

$$S = \begin{pmatrix} 1 & S_{12} \\ S_{21} & S_{22} \end{pmatrix}.$$

By (5.15)

$$\begin{aligned} S_{12} &= p_1, \\ p_1 S_{12} &= 1 + S_{12}, \\ p_2 S_{21} &= S_{22}, \\ p_2 S_{22} &= S_{22} + S_{21}. \end{aligned}$$

These relations hold for $S_{12} = p_1$, $S_{21} = 1$, $S_{22} = p_2$.

Therefore

$$S = \begin{pmatrix} 1 & p_1 \\ 1 & p_2 \end{pmatrix}.$$

By the computation rule of an inverse matrix we have

$$S^{-1} = \frac{1}{p_2 - p_1} \begin{pmatrix} p_2 & -p_1 \\ -1 & 1 \end{pmatrix}.$$

By formula (5.10)

$$x_k = S^{-1} \begin{pmatrix} p_1^k & 0 \\ 0 & p_2^k \end{pmatrix} S \begin{pmatrix} 1 \\ 1 \end{pmatrix} =$$

$$= \frac{1}{-\sqrt{5}} \begin{pmatrix} p_2 & -p_1 \\ -1 & 1 \end{pmatrix} \begin{pmatrix} p_1^k & 0 \\ 0 & p_2^k \end{pmatrix} \begin{pmatrix} p_1+1 \\ p_2+1 \end{pmatrix} =$$

$$= \frac{1}{\sqrt{5}} \begin{pmatrix} -p_2 & p_1 \\ 1 & -1 \end{pmatrix} \begin{pmatrix} p_1^k & 0 \\ 0 & p_2^k \end{pmatrix} \begin{pmatrix} p_1^2 \\ p_2^2 \end{pmatrix} =$$

$$= \frac{1}{\sqrt{5}} \begin{pmatrix} -p_2 p_1^{k+2} + p_1 p_2^{k+2} \\ \\ p_1^{k+2} - p_2^{k+2} \end{pmatrix} = \frac{1}{\sqrt{5}} \begin{pmatrix} p_1^{k+1} - p_2^{k+1} \\ \\ p_1^{k+2} - p_2^{k+2} \end{pmatrix}.$$

This yields a formula for the Fibonacci numbers f_k

$$f_k = \frac{1}{\sqrt{5}} \left[\left(\frac{1+\sqrt{5}}{2} \right)^{k+1} - \left(\frac{1-\sqrt{5}}{2} \right)^{k+1} \right].$$

A careful reader can easily see here a direct analogy with the integration of linear differential equations by means of reducing the system to the Jordan form with the subsequent returning into the initial phase space.

For a linear nonhomogeneous discrete system

$$x_k = A x_{k-1} + f_{k-1} \tag{5.16}$$

we have the following representation of solutions

$$x_k = A^k x_0 + \sum_{j=0}^{k-1} A^{k-j-1} f_j. \tag{5.17}$$

The proof of this formula can be obtained by substitution of the right-hand side of expression (5.17) into the right-hand and left-hand sides of equation (5.16). Notice that, unlike the theory of differential equations, in this case the answer to the question of the existence and uniqueness of solutions of discrete systems

$$x_k = F(x_{k-1}, k-1) \qquad k = 0, 1, \ldots$$

are positive. Consider now a system

$$x_{k+1} = Ax_k + b\xi_k, \qquad (5.18)$$
$$\sigma_k = c^* x_k,$$

where A is a constant $n \times n$-matrix, b and c are constant $n \times m$ and $n \times l$-matrices respectively. We consider ξ_k as an input of a linear block, σ_k as an output, and x_k as a state vector of block at discrete times $k = 0, 1, \ldots$

Definition 5.3. *System* (5.18) *is said to be controllable if for any pair of vectors* $y \in \mathbb{R}^n$, $z \in \mathbb{R}^n$ *there exist a natural number* $N \leq n$ *and a vector sequence* ξ_k *such that a solution* x_k *of system* (5.18) *with the sequence* ξ_k *and the initial data* $x_0 = y$ *satisfies a relation* $x_N = z$.

Theorem 5.2. *System* (5.18) *is controllable if and only if the pair* (A, b) *is controllable.*

Proof. Let us recall that the controllability of the pair (A, b) is equivalent to the following relation

$$\operatorname{rank}(b, Ab, \ldots, A^{n-1}b) = n. \qquad (5.19)$$

Put $N = n$ and apply formula (5.17). Then we obtain

$$x_n = A^n x_0 + \sum_{j=0}^{n-1} A^{k-1-j} b\xi_j. \qquad (5.20)$$

Now we rewrite this equation in the following way

$$b\xi_{n-1} + Ab\xi_{n-2} + \ldots + A^{n-1}b\xi_0 = z - A^n y. \qquad (5.21)$$

Equation (5.21) can be given in the form

$$(b, Ab, \ldots, A^{n-1}b) \begin{pmatrix} \xi_{n-1} \\ \vdots \\ \xi_0 \end{pmatrix} = z - A^n y. \qquad (5.22)$$

The solvability of this linear equation with respect to the unknown matrix

$$\begin{pmatrix} \xi_{n-1} \\ \vdots \\ \xi_0 \end{pmatrix}$$

follows from condition (5.19).

Thus, if the pair (A,b) is controllable, then from (5.22) one can find a sequence of inputs $\xi_0, \ldots, \xi_{n-1}$, which take the state vector x from the state $x_0 = y$ to the state $x_n = z$.

Suppose that (A, b) is not controllable. Then by Theorem 3.2 of Chapter 3 there exists a nonsingular matrix S such that

$$S^{-1}AS = \begin{pmatrix} A_{11} & A_{12} \\ 0 & A_{22} \end{pmatrix}, \qquad S^{-1}b = \begin{pmatrix} b_1 \\ 0 \end{pmatrix}.$$

Let us perform a change of variables $x_k = Sy_k$ in equations (5.18):

$$y_{k+1} = S^{-1}ASy_k + S^{-1}b\xi_k, \qquad \sigma_k = c^*Sy_k. \tag{5.23}$$

Equations for y_k can be given in the following way

$$y_k^{(1)} = A_{11}y_k^{(1)} + A_{12}y_k^{(2)} + b_1\xi_k, \quad y_k^{(2)} = A_{22}y_k^{(2)},$$

where

$$y_k = \begin{pmatrix} y_k^{(1)} \\ \\ y_k^{(2)} \end{pmatrix}.$$

We see that by the input ξ_k it is impossible to transfer the vector $y_k^{(2)}$ from an arbitrary point $y_0^{(2)} = z_1^{(2)}$ to that $y_N^{(2)} = z_2^{(2)}$.

Thus, the fact that the pair (A, b) is not controllable results in that system (5.18) is not controllable too.

As for the continuous system, system (5.18) is said to be observable if a sequence σ_k determines uniquely a sequence x_k.

Here the relation between σ_k and x_k are given by

$$\begin{aligned} &\sigma_0 = c^*x_0, \\ &\sigma_1 = c^*Ax_0 + c^*b\xi_0, \\ &\cdots\cdots\cdots\cdots\cdots\cdots\cdots\cdots \\ &\sigma_n = c^*A^{n-1}x_0 + \sum_{j=0}^{n-1} c^*A^{n-1-j}b\xi_j. \end{aligned}$$

By the above equations the initial data of the block are uniquely determined if and only if

$$\operatorname{rank} \begin{pmatrix} c^* \\ \vdots \\ c^*A^{n-1} \end{pmatrix} = n.$$

Thus we establish the following

Theorem 5.3. *The observability of system* (5.18) *is equivalent to the observability of a pair* (A, c).

For discrete systems we have the theorems similar to those on linear stabilization. Consider one of them.

Let $\xi_k = s^* x_k$. Then by (5.18)

$$x_{k+1} = (A + bs^*)x_k.$$

We recall that under the assumption that $m = 1$ (i.e., b is an n-vector) and (A, b) is controllable the following theorem was obtained, which together with the theorem on stability of a linear homogeneous discrete system allows us to answer the question of stabilization, using the linear feedback of the form $\xi = s^* x$.

Theorem 5.4. *For any polynomial*

$$\psi(p) = p^n + \psi_{n-1} p^{n-1} + \ldots + \psi_0$$

there exists a vector $s \in \mathbb{R}^n$ *such that*

$$\det(pI - A - bs^*) = \psi(p)$$

For discrete systems a Z-transformation is often used. This transformation is similar to the Laplace transformation for continuous systems.

Consider a sequence f_k, which increases as $k \to \infty$ not faster than d^k, where d is any number. We define a Z-transformation of sequence f_k into a function of complex variable z in the following way:

$$Z\{(f_k)\} = F(z) = \sum_{k=0}^{\infty} z^{-k} f_k.$$

Here z lies outside a circle $\{z| \, |z| > d\}$. Note that in the various fields of mathematics there are often used the generating functions of the sequences $G(z) = \sum_{k=0}^{\infty} z^k f_k$. We have $G(z) = F(z^{-1})$.

We see that f_k may be a vector sequence of any dimension.

We apply Z-transformation to both sides of equations (5.18) under the assumption $z_0 = 0$:

$$\sum_{k=0}^{\infty} z^{-k} x_{k+1} = A \sum_{k=0}^{\infty} z^{-k} x_k + b \sum_{k=0}^{\infty} z^{-k} \xi_k,$$

$$\sum_{k=0}^{\infty} z^{-k}\sigma_k = c^* \sum_{k=0}^{\infty} z^{-k}x_k.$$

These equations can be rewritten as

$$\sum_{k=0}^{\infty} z^{-k}\sigma_k = -c^*(A - zI)^{-1}b \sum_{k=0}^{\infty} z^{-k}\xi_k. \tag{5.24}$$

As in the case of continuous system, we call the function $W(z) = c^*(A - zI)^{-1}b$ a transfer function of system (5.18).

A transfer function relates Z-transforms of the input and output of a linear discrete block like that a similar function $W(z)$ relates the Laplace transforms of the input and output for continuous system.

5.3 The discrete phase locked loops for array processors

In modern computers the problems of synchronization the work of processors arise. Since in the array processors the clock skew may be significant [25], it may lead to an incorrect work of parallel algorithms.

The problem of a clock skew in high-speed systems is so much important that a modern VLSI are often supplied by several phase locked loops, placed on one chip [45].

In this case for creating a distributed system of generators [9] the phase locked loops can be used.

We consider the simplest discrete phase locked loop without filters. The main principles of an operation of a continuous phase locked loop was already considered and the equations, describing this work, was obtained in (§4.2). Here the analogue of equation (4.38) is an equation

$$\theta_{k+1} - \theta_k + \alpha F(\theta_k) = \Gamma, \tag{5.25}$$

where α is some number.

We consider further the sinusoidal characteristics of a phase detector, $F(\theta) = \sin\theta$.

The purpose of control is an elimination of a clock skew for almost all initial data θ_0:

$$\lim_{k\to+\infty} \theta_k = 2j\pi. \tag{5.26}$$

From relations (5.26) and (5.25) it follows that $\Gamma = 0$.

We arrive at an equation

$$\theta_{k+1} - \theta_k + \alpha \sin \theta_k = 0. \tag{5.27}$$

Without loss of generality it can be assumed that $\alpha > 0$. The following solutions

$$\theta = 2j\pi, \tag{5.28}$$

$$\theta = (2j+1)\pi \tag{5.29}$$

are equilibria of this equation. Now we consider the linearized form of equation (5.27) in a neighborhood of these equilibria

$$\theta_{k+1} - \theta_k + \alpha(\theta_k - 2j\pi) = 0, \tag{5.30}$$

$$\theta_{k+1} - \theta_k - \alpha(\theta_k - (2j+1)\pi) = 0. \tag{5.31}$$

From Theorem 5.1 it follows that solutions (5.28) of equation (5.30) are asymptotically stable if

$$\alpha < 2. \tag{5.32}$$

Solutions (5.29) of equations (5.31) are unstable.

These properties of solutions (5.28) and (5.29) are also valid for system (5.27) under the condition $\alpha \in (0, 2)$.

We denote by α_1 a root of the equation

$$\sqrt{\alpha^2 - 1} = \pi + \arccos \frac{1}{\alpha}.$$

We show now that any solution of equation (5.27) with initial data $\theta_0 \neq (2j+1)\pi$ under condition (5.32) tends to one of stable equilibria.

Proposition 5.1. *If* $\alpha < \alpha_1$ *and* $\theta_0 \in [-\pi, \pi]$, *then* $\theta_k(k) \in [-\pi, \pi]$ *for all* $k = 1, 2, \ldots$.

Proof. The function $g(\theta) = \theta - \alpha \sin \theta$ satisfies the following inequalities

$$g(\theta) \geq \theta \geq -\pi, \quad \forall \theta \in [-\pi, 0],$$

$$g(\theta) \leq \max_{[-\pi,0]} g(\theta) = \sqrt{\alpha^2 - 1} - \left(\arccos \frac{1}{\alpha}\right) \leq \pi, \quad \forall \theta \in [-\pi, 0],$$

$$g(\theta) \leq \theta \leq \pi, \quad \forall \theta \in [0, \pi],$$

$$g(\theta) \geq \min_{[0,\pi]} g(\theta) = \left(\arccos \frac{1}{\alpha}\right) - \sqrt{\alpha^2 - 1} \geq -\pi, \quad \forall \theta \in [0, \pi].$$

From the above inequalities and from the inclusion $\theta_0 \in [-\pi, \pi]$ the inclusion $\theta_k \in [-\pi, \pi]$ follows.

Proposition 5.1 gives a sharp estimate of mapping the segment $[-\pi, \pi]$ into itself. Really, arguing as in the proof of Proposition 5.1 we obtain that for $\alpha > \alpha_1$ there exists a point $\theta_0 \in [-\pi, \pi]$ such that

$$|\theta_0 - \alpha \sin \theta_0| > \pi.$$

Obviously, $\alpha_1 > 2$. Therefore the proposition is valid under condition (5.32).

If $\theta \in \left(-\frac{\pi}{2}, \frac{\pi}{2}\right)$, then for $\alpha \in (0, 2)$ we have

$$\left|(\theta - \alpha \sin \theta)'\right| < 1. \tag{5.33}$$

On the set $[\pi/2, \pi]$ the following inequality

$$\theta - \alpha \sin \theta > -\frac{\pi}{2} \tag{5.34}$$

is satisfied and on the set $[-\pi, -\pi/2]$ an estimate holds

$$\theta - \alpha \sin \theta < \frac{\pi}{2}. \tag{5.35}$$

From inequalities (5.34) and (5.35) it follows that for any solution θ_k of equation (5.27) with initial data $\theta_0 \in (0, \pi)$ either for all k the estimate $0 < \theta_{k+1} < \theta_k$ holds or for some k the inclusion

$$\theta_k \in \left(-\frac{\pi}{2}\ \frac{\pi}{2}\right) \tag{5.36}$$

is valid. From the inequalities $0 < \theta_{k+1} < \theta_k$ for all k it follows directly that

$$\lim_{k \to +\infty} \theta_k = 0. \tag{5.37}$$

Inclusion (5.36) and estimate (5.33) imply also relation (5.37).

The case $\theta_0 \in (-\pi, 0)$ may be considered in the same way.

Thus the following result is proved.

Proposition 5.2. *For any solution of equation* (5.27) *with* $\alpha \in (0,2)$ *under the initial data* $\theta_0 \neq (2j+1)\pi$ *there exists an integer number* N *such that*

$$\lim_{k \to +\infty} \theta_k = 2N\pi.$$

Below the results of a computer simulation for equation (5.27) for $\alpha \in (2, \alpha_1)$ [30] are given. For $\alpha \in (2,3]$ there exists an asymptotically stable limit cycle, which is symmetric with respect to $\theta = 0$, with period 2 and with a domain of attraction, namely

$$(-\pi, 0) \cup (0, \pi).$$

For all initial data $\theta_0 = -3; -2.9; \ldots - 0.1; 0.1; 0.2; \ldots 2.9; 3$ after 100 iterations for $\alpha = 2.2; 2.3; \ldots; 3$ such limit cycles were obtained. They are given in Table 1.

Table 1

α	value of cycle	
2.2	0.7489866426	-0.7489866426
2.3	0.9028834041	-0.9028834041
2.4	1.026738291	-1.026738291
2.5	1.131102585	-1.131102585
2.6	1.221496214	-1.221496214
2.7	1.301256148	-1.301256148
2.8	1.372589846	-1.372589846
2.9	1.437050573	-1.437050573
3	1.495781568	-1.495781568

Thus, for $\alpha = 2$ the first bifurcation occurs. The globally asymptotic stability of a stationary set vanishes and a unique on $[-\pi, \pi]$ globally asymptotically stable cycle with period 2 is generated. The amplitude of this cycle increases with α.

In some neighborhood of the parameter $\alpha = 3.1$ the second bifurcation occurs: the cycle loses its stability and two locally stable nonsymmetrical cycles of period 2 occur. In the computer simulation for $\alpha = 3.1$ the instability occurs in the following way.

Here, as in the previous cases, after 100 iterations for the initial values $\theta_0 = -3; -2.9; \ldots; -0.1; 0.1; \ldots 3$ we observe the transition into a certain neighborhood

$$\theta_k \in (-1.5496 - \varepsilon, -1.5496 + \varepsilon),$$
$$\theta_{k+1} \in (1.5496 - \varepsilon, 1.5496 + \varepsilon),$$

or

$$\theta_k \in (1.5496 - \varepsilon, 1.5496 + \varepsilon),$$
$$\theta_{k+1} \in (-1.5496 - \varepsilon, -1.5496 + \varepsilon)$$

where $k \geq 100$, $0 < \varepsilon < 0.0002$. For example, for $\theta_0 = 2.5$ we have the following sequence

$$\begin{aligned}
\theta_{101} &= -1.549501645\\
\theta_{102} &= 1.549795512\\
\theta_{103} &= -1.549520909\\
\theta_{104} &= 1.549775220\\
\theta_{105} &= -1.549537730\\
\theta_{106} &= 1.549761807.
\end{aligned}$$

Here the irregularity in the last significant digit is typical and it is a result of the instability of cycle.

For $\alpha = 3.2$ there exists a transition (depending on initial data $\theta_0 = -3; \ldots; -0.1; 0.1; \ldots; 3$) to one of nonsymmetrical locally stable cycles of period 2. The values of these cycles are the following

$$-1.379442769\,; \quad 1.762149884$$

and

$$-1.762149884\,; \quad 1.379442769.$$

We obtain the similar result for $\alpha = 3.3$. Here the values of locally stable cycles are the following

$$-1.259697452\,; \quad 1.881895201$$

and

$$-1.881895201\,; \quad 1.259697452$$

For $\alpha = 3.4$ the instability occurs. Here we can observe the following values:

$$-1.178465\,; \quad 1.963186142$$

and

$$-1.963186142\,; \quad 1.178465.$$

For $\alpha = 3.5$ we have locally stable cycles with period 4:

$$\begin{array}{ll} -0.9994532\,; & 1.9446608; \\ -1.3135694\,; & 2.0712776 \end{array}$$

and

$$\begin{array}{ll} 0.9994532\,; & -1.9446608; \\ 1.3135694\,; & -2.0712776. \end{array}$$

For $\alpha = 3.52$ we see the transient processes, which yields the locally stable cycle $\theta_k^{(1)}$ of period 8:

$$\begin{array}{ll} 1.003657629\,; & -1.965256723\,; \\ 1.284421617\,; & -2.092223694\,; \\ 0.960000216\,; & -1.923554541\,; \\ 1.379695669\,; & -2.076225446 \end{array}$$

or a cycle to be symmetric to it: $\theta_k^{(2)} = -\theta_k^{(1)}$.

For $\alpha = 3.7$ we see chaotic dynamics without periodic or quasiperiodic trajectories. Here we observe a local instability.

Thus, there exists the transition to chaos via a sequence of period doubling bifurcations [42]. Unlike the classic period doubling bifurcations in logistic mappings [42] the interesting property of the system considered is the fact that before the period doubling bifurcation, from one globally asymptotically stable cycle of period two there occur two locally stable cycles of period two.

Various aspects, concerning the development of the theory of discrete dynamical systems and their applications, can be found in [6, 21, 22, 24, 43, 44, 46].

Chapter 6

The Aizerman conjecture. The Popov method

In 1949 M.A. Aizerman, the notorious specialist in the control theory, has stated a conjecture [1], which stimulated the development of new mathematical methods in studying nonlinear differential equations. At present these methods go far beyond the control theory and are used in solving many applied problems.

Let us consider a system

$$\begin{aligned} \frac{dx}{dt} &= Ax + b\varphi(\sigma), \\ \sigma &= c^* x, \end{aligned} \tag{6.1}$$

where A is a constant $n \times n$-matrix, b and c are constant n-dimensional vectors, $\varphi(\sigma)$ is a continuous function.

In addition, we consider the following linear systems

$$\begin{aligned} \frac{dy}{dt} &= Ay + b\mu\sigma, \\ \sigma &= c^* x \end{aligned} \tag{6.2}$$

and suppose that for

$$\alpha < \mu < \beta \tag{6.3}$$

all the solutions of systems (6.2) tend to zero as $t \to +\infty$.

The question arises whether the zero solution of system (6.1) is globally asymptotically stable if the following condition

$$\alpha < \varphi(\sigma)/\sigma < \beta \tag{6.4}$$

is satisfied for all $\sigma \neq 0$.

We say that the zero solution of system (6.1) is globally asymptotically stable (or system (6.1) is globally stable) if this zero solution is Lyapunov stable and any solution of system (6.1) tends to zero as $t \to +\infty$.

M.A. Aizerman conjectured that the answer to this question is always positive.

In the papers of N.N. Krasovskii, V.A. Pliss, E. Noldus, and G.A. Leonov the classes of nonlinear systems were given for which the Aizerman conjecture is not true. That is, inequality (6.4) is satisfied, all the linear systems (6.2) with μ, satisfying (6.3), are stable but system (6.1) has the solutions that do not tend to zero as $t \to +\infty$. These results are considered in [28,29].

Now we demonstrate a method to obtain the criteria of global stability. The method was suggested by V.M. Popov [37] in 1958. The investigations of V.M. Popov were stimulated in part due to the Aizerman problem.

Consider system (6.1) under the additional assumption that A is a stable matrix, i.e., all its eigenvalues have negative real parts. Now we change slightly condition (6.4). Suppose, the following inequality holds

$$0 \leq \varphi(\sigma)/\sigma \leq k, \qquad \forall \sigma \neq 0, \tag{6.5}$$

where k is any number.

As in Chapter 2, consider a transfer function of the linear part of system (6.1)

$$W(p) = c^*(A - pI)^{-1}b, \qquad p \in \mathbb{C}.$$

The Popov theorem. *Suppose that there exists a number θ such that for all real ω inequalities hold*

$$\begin{aligned} &\frac{1}{k} + \mathrm{Re}\left[(1+\theta i\omega)W(i\omega)\right] > 0, \\ &\lim_{\omega \to +\infty}\left[\frac{1}{k} + \mathrm{Re}\left[(1+\theta iw)W(i\omega)\right]\right] > 0. \end{aligned} \tag{6.6}$$

Then system (6.1) is globally asymptotically stable.

Before proving the theorem, let us recall the definition of the Fourier transformation and some of its properties.

In Chapter 2 two different definitions of the Fourier transformation were cited. For an absolutely integrable and piecewise continuous on $(-\infty, +\infty)$ (and for F on $(0, +\infty)$) function $f(t)$ they are given by the following way

$$\mathcal{F}(f(t)) = \frac{1}{\sqrt{2\pi}} \int\limits_{-\infty}^{+\infty} e^{i\omega t} f(t)\, dt, \quad F(f(t)) = \int\limits_{0}^{+\infty} e^{-i\omega t} f(t)\, dt.$$

For the Fourier transformation the following inversion formula is well known

$$f(t) = \frac{1}{2\pi} \int\limits_{-\infty}^{+\infty} \left(\int\limits_{-\infty}^{+\infty} e^{i\omega(\tau - t)} f(\tau)\, d\tau \right) d\omega, \tag{6.7}$$

which is proved in the second volume of the text-book [14].

Here the exterior integral is understood in the sense of the principal value, i.e., as the following limit

$$\lim_{M\to\infty} \int\limits_{-M}^{M} (\ldots)\, d\omega.$$

By (6.7) for the absolutely integrable product $f(t)g(t)$ of the absolutely integrable and piecewise continuous functions $f(t)$ and $g(t)$, the following relations hold

$$\int\limits_{-\infty}^{+\infty} f(t)g(t)\, dt = \int\limits_{-\infty}^{+\infty} \frac{g(t)}{2\pi} \left(\int\limits_{-\infty}^{+\infty} \left(\int\limits_{-\infty}^{+\infty} e^{i\omega(\tau - t)} f(\tau)\, d\tau \right) d\omega \right) dt =$$

$$= \int\limits_{-\infty}^{+\infty} \left(\frac{1}{\sqrt{2\pi}} \int\limits_{-\infty}^{+\infty} e^{-i\omega t} g(t)\, dt \right) \left(\frac{1}{\sqrt{2\pi}} \int\limits_{-\infty}^{+\infty} e^{i\omega\tau} f(\tau)\, d\tau \right) d\omega$$

Hence we have

$$\int\limits_{-\infty}^{+\infty} f(t)g(t)\, dt = \int\limits_{-\infty}^{+\infty} \mathcal{F}(f(t))\, \overline{\mathcal{F}(g(t))}\, d\omega. \tag{6.8}$$

Relation (6.8) demonstrates a very important property of the Fourier transformation, namely the transformation is a unitary operator in the space of square summable functions $L_2(-\infty, +\infty)$. In other words, if for a set of square summable functions f and g a scalar product

$$(f, g) = \int_{-\infty}^{+\infty} f(t)g(t)\,dt$$

is well defined, then the operator $\mathcal{F}$ acts on a set of such functions, preserving a scalar product.

From (6.8) it follows that for the Fourier transformation F the relation

$$\int_{0}^{+\infty} f(t)g(t)\,dt = \frac{1}{2\pi}\int_{-\infty}^{+\infty} F(f(t))\overline{F(g(t))}\,d\omega \tag{6.9}$$

is satisfied. We also recall that in Chapter 2 the following properties

$$F(\dot{f}(t)) = i\omega\, F(f(t)) - f(0), \tag{6.10}$$

$$F\Big(\int_{0}^{t} f(t-\tau)g(\tau)\,d\tau\Big) = F(f(t))\;F(g(t)) \tag{6.11}$$

were proved (Propositions 2.1 and 2.2).

Now we can proceed to the proof of the Popov theorem.

Note that if inequalities (6.6) are valid, then there exists a number $\varepsilon > 0$ such that

$$\frac{1}{k} - \varepsilon + \operatorname{Re}\big[(1+\theta i\omega)W(i\omega)\big] > 0, \quad \forall\omega \in \mathbb{R}^1. \tag{6.12}$$

We notice also that, having performed the change of nonlinearity $\varphi(\sigma) = k\sigma - \psi(\sigma)$ and using in place of system (6.1) the following equations

$$\dot{x} = (A + bc^*k)x - b\psi(\sigma), \qquad \sigma = c^*x, \tag{6.13}$$

it is possible to reduce condition (6.6) to a formally more restrictive condition (6.6) with $\theta \geq 0$. Really, in this case system (6.13) have a transfer function

$$U(p) = \frac{-W(p)}{1 + kW(p)}$$

and $\psi(\sigma)$ satisfied condition (6.5):

$$0 \leq \psi(\sigma)/\sigma \leq k, \qquad \forall \sigma \neq 0.$$

Inequality (6.6) for system (6.13) takes the form

$$\frac{1}{k}+\mathrm{Re}\left[(1+\theta i\omega)U(i\omega)\right]=$$
$$=\mathrm{Re}\,\frac{(1+kW(i\omega)-k\theta i\omega W(i\omega)-k^2\theta i\omega|W(i\omega)|^2)}{k|1+kW(i\omega)|^2}=$$

$$=\frac{\dfrac{1}{k}+\mathrm{Re}\,((1-\theta i\omega)W(i\omega))}{|1+kW(i\omega)|^2}>0.$$

Thus, without loss of generality we can assume that $\theta \geq 0$.

Lemma 6.1. *If inequality* (6.12) *is satisfied, then for any number* $T > 0$ *the inequality*

$$\int_0^T \left[\varphi(\sigma(t))\left(\sigma(t)-\frac{1}{k}\varphi(\sigma(t))\right)+\varepsilon\varphi(\sigma(t))^2+\right.$$
$$\left.+\theta\varphi(\sigma(t))\dot{\sigma}(t)\right]dt \leq \int_0^T \varphi(\sigma(t))\big(\alpha(t)+\theta\dot{\alpha}(t)\big)dt. \tag{6.14}$$

is valid.

P r o o f. Consider functions

$$\varphi_T(t)=\begin{cases}\varphi(\sigma(t)) & \text{for} \quad t\in[0,T],\\ 0 & \text{for} \quad T>0,\end{cases}$$

$$\sigma_T(t)=\alpha(t)+\int_0^t \gamma(t-\tau)\varphi_T(\tau)\,d\tau.$$

Here T is a certain positive number,

$$\alpha(t)=c^*e^{At}x_0, \qquad \gamma(t)=c^*e^{At}b.$$

Sometimes the function $\varphi_T(t)$ is called a cut-off function of $\varphi(\sigma(t))$.

Obviously, the following relations hold

$$\int_0^T \left[\varphi(\sigma(t))\left(\sigma(t)-\frac{1}{k}\,\varphi(\sigma(t))\right)+\varepsilon\varphi(\sigma(t))^2+\theta\varphi(\sigma(t))\dot\sigma(t)\right]dt=$$

$$=\int_0^{+\infty}\left[\varphi_T(t)\left(\sigma_T(t)-\frac{1}{k}\,\varphi_T(t)\right)+\varepsilon\varphi_T(t)^2+\theta\varphi_T(t)\dot\sigma_T(t)\right]dt=$$

$$=\int_0^{+\infty}\varphi_T(t)\Bigg[\int_0^t\gamma(t-\tau)\varphi_T(\tau)\,d\tau+\left(\varepsilon-\frac{1}{k}\right)\varphi_T(t)+ \tag{6.15}$$

$$+\theta\,\frac{d}{dt}\left(\int_0^t\gamma(t-\tau)\varphi_T(\tau)\,d\tau\right)\Bigg]dt+\int_0^{+\infty}\varphi_T(t)\big(\alpha(t)+\theta\dot\alpha(t)\big)\,dt.$$

By (6.9)–(6.11) we obtain

$$\int_0^{+\infty}\varphi_T(t)\Bigg[\int_0^t\gamma(t-\tau)\varphi_T(\tau)\,d\tau+\left(\varepsilon-\frac{1}{k}\right)\varphi_T(t)+$$

$$+\theta\,\frac{d}{dt}\left(\int_0^t\gamma(t-\tau)\varphi_T(\tau)\,d\tau\right)\Bigg]dt=$$

$$=\frac{1}{2\pi}\int_{-\infty}^{+\infty}\overline{F(\varphi_T(t))}\Bigg[F(\gamma(t))\,F(\varphi_T(t))+\left(\varepsilon-\frac{1}{k}\right)F(\varphi_T(t))+$$

$$+\theta i\omega F(\gamma(t))\,F(\varphi_T(t))\Bigg]d\omega=$$

$$=\frac{1}{2\pi}\int_{-\infty}^{+\infty}\left[-(1+\theta i\omega)W(i\omega)+\left(\varepsilon-\frac{1}{k}\right)\right]|F(\varphi_T(t))|^2d\omega.$$

We see that if inequality (6.12) is satisfied, then the quantity

$$\int_0^{+\infty} \varphi_T(t)\Big[\int_0^t \gamma(t-\tau)\varphi_T(\tau)\,d\tau + \left(\varepsilon - \frac{1}{k}\right)\varphi_T(t) +$$

$$+\theta\,\frac{d}{dt}\left(\int_0^t \gamma(t-\tau)\varphi_T(\tau)d\tau\right)\Big]\,dt$$

is negative. Hence relations (6.15) yields inequality (6.14).

Lemma 6.2. *If inequality* (6.12) *is valid for* $\theta > 0$, *then there exists a number* q *such that*

$$\int_0^{+\infty} \varphi(\sigma(t))^2 dt \leq q|x_0|^2. \tag{6.16}$$

Proof. From condition (6.5) and $\theta \geq 0$ by Lemma 6.1 it follows that

$$\varepsilon \int_0^T \varphi(\sigma(t))^2 dt \leq -\theta \int_{\sigma(0)}^{\sigma(T)} \varphi(\sigma)\,d\sigma + \int_0^T \varphi(\sigma(t))\left(\alpha(t) + \theta\dot{\alpha}(t)\right) dt.$$

By using the inequality

$$uv \leq \frac{\varepsilon}{2}u^2 + \frac{1}{2\varepsilon}v^2, \qquad \forall\, u,\ \forall\, v$$

and the fact that

$$\int_0^\sigma \varphi(\sigma)\,d\sigma \geq 0, \quad \forall\,\sigma$$

(the last inequality results from condition (6.5)), we obtain the following estimate

$$\frac{\varepsilon}{2}\int_0^T \varphi(\sigma(t))^2 dt \le \theta \int_0^{\sigma(0)} \varphi(\sigma)\,d\sigma + \frac{1}{2\varepsilon}\int_0^T (c^* e^{At}x_0 + \theta c^* A e^{At} x_0)^2 dt \le$$

$$\le k|c|^2|x_0|^2 + \frac{1}{2\varepsilon}\int_0^T (c^* e^{At}x_0 + \theta c^* A e^{At} x_0)^2 dt.$$

From the above and stability of the matrix A the statement of Lemma 6.2 follows.

Consider now a linear system

$$\dot{x} = Ax + f(t), \qquad x \in \mathbb{R}^n, \tag{6.17}$$

where A is a constant stable matrix, $f(t)$ is a continuous vector function.

Lemma 6.3. *Suppose, for some number α the following inequality holds*

$$\int_0^{+\infty} |f(t)|^2 dt < \alpha. \tag{6.18}$$

Then there exists a number β such that the inequality

$$|x(t)|^2 \le \beta(|x(0)|^2 + \alpha) \tag{6.19}$$

is satisfied. In addition, the following relations hold

$$\int_0^{+\infty} |x(t)|^2 dt < +\infty, \qquad \lim_{t\to+\infty} x(t) = 0. \tag{6.20}$$

Proof. Recall (see Chapter 1) that there exists a positively definite matrix H such that

$$A^*H + HA = -I.$$

Therefore for the function $W(t) = x(t)^* H x(t)$ we have the following estimate

$$\begin{aligned} W(x(t))^\bullet &= -|x(t)|^2 + 2x^* H f(t) \le \\ &\le -|x(t)|^2 + |2H|\,|x(t)|\,|f(t)| \le -\frac{1}{2}|x(t)|^2 + 2|H|^2|f(t)|^2. \end{aligned} \tag{6.21}$$

Hence

$$x(t)^* H x(t) \leq x(0)^* H x(0) + 2\alpha |H|^2. \tag{6.22}$$

From the last inequality and the positive definiteness of matrix H the existence of a number β follows such that inequality (6.19) is satisfied.

Integrating both sides of inequality (6.21) from 0 to t, we obtain

$$W(x(t)) - W(x(0)) + \frac{1}{2} \int_0^t |x(\tau)|^2 d\tau \leq 2\alpha |H|^2.$$

By (6.22) we have

$$\int_0^{+\infty} |x(t)|^2 dt < \infty. \tag{6.23}$$

Now we consider an integral

$$\int_0^t x(\tau)^* \dot{x}(\tau)\, d\tau = \frac{1}{2}|x(t)|^2 - \frac{1}{2}|x(0)|^2. \tag{6.24}$$

It is obvious that the following relations hold

$$\int_0^t |x(\tau)\dot{x}(\tau)|\, d\tau \leq \int_0^t |x(\tau)|^2 d\tau + \int_0^t |\dot{x}(\tau)|^2 d\tau =$$

$$= \left(\int_0^t |x(\tau)|^2 d\tau + \int_0^t |Ax(\tau) + f(\tau)|^2 dt \right) \leq$$

$$\leq \left(\int_0^t |x(\tau)|^2 d\tau + \int_0^t \left(|A|\, |x(\tau)| + |f(\tau)|\right)^2 dt \right) \leq$$

$$\leq \left((1 + 2|A|^2) \int_0^t |x(\tau)|^2 d\tau + 2 \int_0^t |f(\tau)|^2 dt \right).$$

By (6.18) and (6.23) the integral

$$\int_0^t x(\tau)^* \dot{x}(\tau)\, dt$$

converges absolutely. It means that there exists a limit

$$\lim_{t\to+\infty} \int_0^t x(\tau)^* \dot{x}(\tau)\, d\tau.$$

Then by (6.24) there exists a limit

$$\lim_{t\to+\infty} |x(t)|.$$

From (6.23) it follows that

$$\lim_{t\to+\infty} |x(t)| = 0.$$

The proof of Lemma 6.3 is completed.

Lemmas 6.2 and 6.3 imply that system (6.1) is globally asymptotically stable.

In conclusion we note that in the case $\theta < 0$ it is necessary to make use of the change of variables, mentioned above, and to pass then to the case $\theta \geq 0$.

A geometric interpretation of the Popov theorem. We consider a hodograph of the modified frequency response $X(\omega) = \operatorname{Re} W(i\omega)$, $Y(\omega) = \omega \operatorname{Im} W(i\omega)$ (Fig. 6.1).

If through the point $X = -1/k$, $Y = 0$ it is possible to draw the line $X - \theta Y = -1/k$ such that the total hodograph $\{X(\omega), Y(\omega)\}$ is placed to the right of this straight line, then assumption (6.6) of the Popov theorem is satisfied.

It is interesting to compare this result with the Nyquist criterion, which is a necessary and sufficient condition of stability of linear systems. For linear systems (6.1) with $\varphi(\sigma) = \mu\sigma$ to be stable in the case that $\mu \in (0, k)$ and the matrix A is stable, it is necessary and sufficient for the hodograph $X(\omega)$, $Y(\omega)$ to be not intersect the "forbidden ray" $\{X, Y \mid X \in (-\infty, -1/k), Y = 0\}$.

The Popov frequency condition is more rigorous. A "forbidden zone" is a half-plane to the left of the line $X - \theta Y = -1/k$, where θ is a running parameter. In this case for two-dimensional systems it is possible to show that if a maximal stability sector, obtained by means of the Nyquist criterion, takes the form $[0, k_0)$, then for any $k < k_0$ there exists a parameter θ such that inequalities (6.6) are satisfied.

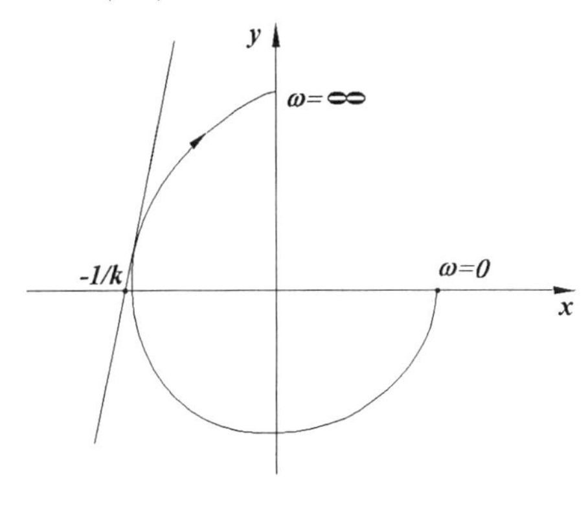

Fig. 6.1

Now we shall show the way in which the Popov theorem can be applied to the study of the Aizerman conjecture. Consider the two-dimensional systems (1) in the case that the linear stability sector (6.3) is finite. This implies that for $\mu = \alpha$ or $\mu = \beta$ the matrix $A + \mu bc^*$ has eigenvalues on the imaginary axis. In this case, at least, one of the matrices $A + \alpha bc^*$ and $A + \beta bc^*$ is nondegenerate.

Show that the Aizerman conjecture has a positive solution for the following class of nonlinearities

$$\alpha + \varepsilon \leq \varphi(\sigma)/\sigma \leq \beta - \varepsilon, \quad \forall \sigma \neq 0. \tag{6.25}$$

Here ε is any positive number.

Without loss of generality we may consider the case that

$$0 \leq \varphi(\sigma)/\sigma \leq K, \quad \forall \sigma \neq 0$$

and

$$W(p) = \frac{\rho p + \nu}{p^2 + \delta p + \gamma},$$

where γ and δ are positive numbers, δ is small, $\rho \geq 0$. In the sequel, we assume that $\rho > 0$.

Suppose also that in (6.6) we have

$$\theta = \frac{\nu}{\rho\gamma}.$$

In this case we obtain

$$\frac{\nu}{\gamma} + O(\delta) \leq \operatorname{Re}\left[(1+\theta i\omega)W(i\omega)\right] \leq \frac{\nu}{\gamma} + O(\delta).$$

Thus, if $\nu > 0$, then for small δ inequalities (6.6) are satisfied for all $K > 0$. If $\nu = 0$, then inequalities (6.6) are satisfied for $K = K(\delta)$. In this case we have $K(\delta) \to +\infty$ as $\delta \to 0$. If $\nu < 0$, then inequalities (6.6) holds for

$$K < -\gamma/\nu + O(\delta).$$

We can easily see that for $\nu < 0$ and $\delta = 0$ a sector

$$0 < \mu < -\gamma/\nu$$

is a maximal sector of linear stability (6.3). The Popov theorem implies that the Aizerman conjecture is valid in the class of functions (6.25), where

$$\alpha = 0, \quad \beta = -\gamma/\nu.$$

Globally stable systems (6.1) with nonlinearities, satisfying condition (6.5), are often called absolutely stable.

An additional information on the theory of absolute stability and various applications of the Popov method an interested reader can find in the books [17, 28, 29, 37].

Bibliography

[1] M.A. Aiserman, (1949) On one problem concerning global stability of dynamical systems [in russian]// Uspehi matematicheskih nauk, Vol.4 No. 4, pp. 187–188.

[2] A.A. Andronov, A.A. Witt, and S.E. Khaikin, (1965) Theory of Oscillation. Pergamon Press.

[3] V.I.Arnold, (1991) Gewöhnliche Differentialgleichungen. Deutscher Verlag der Wissenschaften.

[4] V.I.Arnold, (1978) Mathematical methods of the classical mechanics. Springer.

[5] K.J. Åström, (1999) Automatic control — the hidden technology// Adv. Control Highlights of ECC'99. Springer.

[6] K.J. Åström, B. Wittenmark, (1984) Computer controlled systems. Theory and design. Prentice-Hall.

[7] E.A. Barabashin and V.A. Tabueva, (1969) Dynamical systems with cylindrical phase space [in russian], nauka.

[8] M.M. Botvinnik, (1950) Excitation regulators and static stability of synchronous machine [in russian], Gosenergoizdat.

[9] T.H.Cormen, C.E.Leiserson and R.L.Rivest, (1990) Introduction to algoritms. MIT Press.

[10] A.J. Dirksen, (1979) Microcomputers. Kluwer.

[11] J.J. D'Azzo , C.H. Houpis, (1995) Linear control systems. Analysis and design, Mc Graw-Hill.

[12] M. Driels, (1996) Linear control systems engineering, Mc Graw-Hill.

[13] A.F. Filippov, (1985) Differential equations with discontinuous right hand side [in russian], Nauka.

[14] G.M. Fichtenholz, (1964) Calculus. Vol. I, II [in russian], Nauka.

[15] A.L.Fradkov, A.Yu.Pogromskiy, (1999) Introduction to control of oscillations and chaos. World Scientific.

[16] F.R. Gantmacher, (1959) Theory of matrices. Chelsea.

[17] A.Kh. Gelig, G.A. Leonov and V.A. Jakubovich, (1978) Stability of nonlinear

systems with nonunique equilibrium state [in russian], Nauka.
[18] J. Golten, A. Verwer, (1991) Control system. Design and simulation. Mc Graw-Hill.
[19] N.A. Gubar', (1961) Investigation of some pisewise-linear dynamical system of third order [in russian]// Prikladnaja matematika i mekhanika, Vol. 25, No. 6, pp. 1011—1023.
[20] A.Isidori, (1995) Nonlinear control systems, Springer.
[21] E.Jury, (1958) Sampled-data control systems. John Wiley.
[22] E.Jury, (1964) Theory and application of the Z-transform method. John Wiley.
[23] R.E.Kalman, Y.C.Ho, K.S.Narendra, (1963) Controllability of linear dynamical systems. In contribution to differential equations. P.189-213. John Wiley.
[24] A.Ya. Kosyakin, B.M. Shamrikov, (1983) Oscillation in digital systems [in russian], Nauka.
[25] S.Y.Kung, (1988) VLSI Array processors. Prentice Hall.
[26] S. Lefschetz, (1965) Stability of nonlinear control systems. Academic Press.
[27] G.A. Leonov, (2000) The Brockett stabilization problem// Proceedings of International Conference Control of Oscillations and Chaos. St.Petersburg. P.38–39.
[28] G.A. Leonov, D.V. Ponomarenko, and V.B. Smirnova, (1996) Frequency-domain methods for nonlinear analysis. Theory and applications. World Scientific.
[29] G.A. Leonov, I.M. Burkin, and A.I. Shepelyavyi, (1996) Frequency methods in oscillation theory. Kluwer.
[30] G.A. Leonov and S.M. Seledzhji, (2001) Global stability of phase locked loops. Vestnik St.Petersburg University. Math., No.2, P.67-90.
[31] W.C. Lindsey, (1972) Synchronization systems in communication and control. Prentice-Hall.
[32] Yu.A. Mitropol'skii, (1971) The averaging method in the nonlinear mechanics. Naukova Dumka, [in russian].
[33] L. Morean, D. Aeyels, (1999) Stabilization by means of periodic output feedback// Proceedings of conference of decision and control. Phoenix, Arizona USA. P.108–109.
[34] R.C. Nelson, (1998) Flight stability and automatic control, Mc Graw-Hill.
[35] H. Nijmeijer and A.J. Van der Schaft, (1996) Nonlinear dynamical control systems. Springer.
[36] Open problems in mathematical systems and control theory. (1999) Springer.
[37] V.M. Popov, (1973) Hyperstability of control systems. Springer.
[38] B.V. Rauschenbach and E.N. Tokar', (1974) Orientation control for spacecrafts [in russian], Nauka.
[39] C.E. Rohrs, J.L. Melsa, D.G. Schultz, (1993) Linear control systems, Mc Graw-Hill.
[40] V.V. Shakhgil'dyan and A.A. Lyakhovkin, (1972) Phase locked loops. Svjaz' [in russian].

[41] A.V. Shcheglyaev and S.G. Smil'nitskii, (1962) Regulation of steam turbines [in russian], Gosenergoizdat.
[42] H.G. Schuster, (1984) Deterministic chaos. An introduction. Physik-Verlag.
[43] Ya.Z. Tsypkin, (1958) Theory of sampled-data control systems. [in russian], Nauka.
[44] Ya.Z. Tsypkin, (1977) Basic theory of control systems [in russian], Nauka.
[45] E.P. Ugryumov, (2000) Digital systems, [in russian], bhv.
[46] R.J. Vaccaro, (1995) Digital control, Mc Graw-Hill.
[47] V. Volterra, (1931) Lecons sur la theorie mathematique de la lutte pour la vie. Gauthier-Villars.
[48] A.A. Yanko-Trinitskii, (1958) A new method of analysis of synchronous motors [in russian], Gosenergoizdat.

Index